T0122825

How the ThinkPad Changed the World

and Is Shaping the Future

How the ThinkPad Changed the World

and Is Shaping the Future

Arimasa Naitoh and William J. Holstein

Skyhorse Publishing

Visit our website at www.skyhorsepublishing.com.

10 9 8 7 6 5 4 3 2 1

Library of Congress Cataloging-in-Publication Data is available on file.

Jacket design by Shigeyuki Kimura
Jacket photograph by Joel Collins
Unless otherwise indicated, all photos are courtesy of Lenovo.

Print ISBN: 978-1-5107-2499-0
Ebook ISBN: 978-1-5107-2500-3

Printed in the United States of America

Contents

Introduction

This book is about one of the technology world's most successful creators. Yet, outside of Japan and China, few people have ever heard of Arimasa Naitoh, the father of the ThinkPad notebook computer. Naitoh has stayed with the ThinkPad and the team he built in Japan over the course of a quarter of a century, which is an eternity in the evanescent technology scene. The companies that have failed over that time period could fill an enormous electronic graveyard—think of Commodore or RCA or Zenith.

Naitoh and his team overcame the sorts of challenges that often destroy technology icons—complacency, internal divisions, arrogance, or just plain shortsightedness. In the technology world, there is always a crisis, as Intel's Andy Grove famously wrote in *Only the Paranoid Survive*. If you manage yourself as if there is no crisis, you will be surprised. Taking that cue, Naitoh's team stayed focused and determined to win.

In so doing, they created a device that helped transform the world. Quite literally. The first ThinkPad, launched in 1992, was one of the first devices to enable what we today take for granted—the ability to access our documents, pictures, and entertainment at any time from anywhere, all wirelessly. That capability has transformed the way the business world works, how educators teach the young, how science is performed and innovation is undertaken, how music and video are consumed, and many other aspects of modern human life.

The ThinkPad has been to the top of Mount Everest and to the depths of the ocean. Scientists used it to study biodiversity in the canopies of rain forests. The ThinkPad went on the first expedition to travel the entire length of the Blue Nile and Nile River. NASA enthusiastically embraced the machine, and it has served on both the International Space Station and the Mir space station. Because of one of its designs, the "Butterfly," the ThinkPad is even part of the permanent collection at the Museum of Modern Art in New York.

The very concept of a "road warrior," someone who can stay in touch with colleagues and conduct business from anywhere, whether from home or from a distant airport, was enabled by the ThinkPad. Rather than staying late in the office and missing out on quality time with their children, millions of professionals go home to have dinner with their families, only to turn on their notebook computers after the kids are in bed. That work-life balance has been made possible by the ThinkPad and other devices that followed in its wake. If you wish to, you can always be connected.

In contemplating the impact this machine has had, consider one statistic: more than one hundred million ThinkPads have been sold since its introduction, and many more will be sold by the time the ThinkPad reaches its twenty-fifth anniversary in October 2017. Very few other companies have sold devices on that scale—perhaps only Apple and Samsung Electronics, among the companies left standing. Nokia and Blackberry have been blown away.

It's impossible to determine, as I discovered, just how many models of the ThinkPad have been created and introduced. There have been four major generations—the original launched by the 700C, the XTRA series that was unveiled in 2000 and 2001, the X300 that appeared in 2008, and the X1 that came out in 2011. In addition, a separate class of machines with an entirely different form factor—the ThinkPad Yoga and the ThinkPad X1 Yoga—appeared in the 2013-2016 time frame.

An additional complicating factor, in terms of the number of models produced, is that IBM and later Lenovo, which purchased IBM's PC division including ThinkPad in 2005, customized machines for major corporate clients, and there were also different variations for different countries. Screens were twelve-inch, fourteen-inch, and fifteen-inch, all in different models. Is each of these considered a different model? A further complication is that no one kept track of just how many models were developed for possible introduction into the market and how many actually went on sale. The safest answer is that hundreds of models actually went on sale.

Nor does it seem possible to determine precisely how many patents are associated with the ThinkPad. IBM—for many years—filed for the most US patents of any company in the world. Some knowledge of who held which patents seems to have been lost when Lenovo bought the PC division. IBM retained certain patents but allowed others to move to the acquiring company. Lenovo today does not disclose just how many patents it holds on the ThinkPad for competitive reasons. But the number is clearly in the thousands. In researching this book, I met ThinkPad engineers who had forty to sixty patents each. Overall, many hundreds of very creative engineers have contributed to the ThinkPad over the decades. That very easily adds up to thousands of patents.

The story of how the ThinkPad has developed touches on every major technology trend in consumer electronics and communications—the miniaturization of computers (which used to fill large rooms), the first use of batteries to allow the devices to become mobile, the introduction of color screens, the rise of the Internet and wireless capabilities including the Bluetooth radio wave standard (which was created for the ThinkPad), and most recently, the growing use of touch screens to control and access computers. Any one of those technological transformations could disrupt an existing product or the company that makes the product, but Naitoh and his team kept anticipating these changes and incorporating them into the ThinkPad.

As a result, the ThinkPad is, in some ways, ahead of the field, including Apple. Partly thanks to the infusion of new ideas from China, the ThinkPad is no longer just a square box with a red TrackPoint in the middle of the keyboard. The ThinkPad Yoga and the ThinkPad X1 Yoga have 360-degree hinges that allow four screen positions to facilitate watching movies or referring to recipes in the kitchen. This so-called "convertible" segment of the personal computer market, featuring superthin ultrabooks and tablets, is the hottest-selling category.

Naitoh's team also has been the first in the notebook world to adopt organic light-emitting diodes (OLED) on the ThinkPad X1 Yoga. Others are striving to catch up, but Naitoh's team was first in figuring out the tricky issues of managing this breakthrough technology. OLED represents a leap over traditional liquid-crystal display (LCD) screens and offers stunning color and clarity. These new screens are also thinner than LCD screens, which will allow for the size of notebooks to be driven down further.

The X1 Yoga also incorporates a pen that can be used to draw sketches or concepts on the screen. Those images can be captured and transmitted to others. This feature could create a whole new avenue of collaboration among users. It is like using a whiteboard to sketch out new ideas, but the ideas can be shared electronically with others who may be located far away. In short, it is not your grandfather's ThinkPad anymore.

Naitoh's and my book touches on the impact of a technology on society and business rather than focusing exclusively on "inside baseball"—the details inside a company of how a particular innovation occurred. We wrote the book in this way to make it as widely accessible as possible. You do not have to be a technology genius or a business maven to understand the sweep of the story. Yet, technologically sophisticated readers and business school students will find satisfying content. You also do not need to be a specialist in Japanese or Chinese culture. We provide the context when it's necessary.

This book is more than a history. We will take the reader up to the current day when ThinkPad is engaged in a head-to-head battle with Apple products inside large corporations and universities— and increasingly in the broader consumer market. The ThinkPad and MacBook come from opposite ends of the computing spectrum— the ThinkPad has traditionally appealed to serious professionals who must manage documents and spreadsheets whereas Apple products have appealed to users because of their ability to manage music, pictures, and video. But now, in a high-stakes battle, the two are colliding head-on in the middle of the market as devices become ever sleeker and more multifunctional.

Apple and the ThinkPad unit have very different innovation models. Apple has traditionally depended on the visionary leadership of Steve Jobs, who has passed away, to the regret of everyone in the technology world. History teaches us that it is very hard to fill the vacuum left by a visionary leader. It is very hard to come up with products that create a whole new category, as Jobs did with the iPhone and iPad. Think of what Akio Morita did at Sony—he invented the Sony Walkman, a whole new category of product, but then the world changed and Sony, in his absence, was not able to ride the new waves.

Naitoh, who is now sixty-five, also will be leaving the scene when he retires, but he believes that he has created an innovation model that is not based on a single individual. The cross-fertilization among Japanese, Chinese, and American engineers and marketers, all within Lenovo, is just beginning to yield new ideas and new products. Naitoh has created an army of innovators. In our final chapter, Naitoh offers his vision of the future of mobile computing.

Amazingly, Naitoh's side of the ThinkPad story has never been told in the English language. Previous books and magazine articles overlooked Naitoh's role because they were written from the perspective of executives and marketers in the United States. The reality is that while top management at IBM was fighting over bureaucratic turf in the 1990s, Naitoh was willing the ThinkPad to happen—and keep happening. He has displayed raw determination sustained over multiple decades. That has required enormous stamina, both physically and mentally. In this book, he reveals many more personal details about his victories and his struggles than the vast majority of Japanese would ever do.

Naitoh is an unusual Japanese because he has spent so much time in the United States and China. He has learned to adapt to different cultures and sets of expectations in a way that is rare in the business world. With Japanese engineers or suppliers, he can be profoundly Japanese. Yet, with Western managers or counterparts, he can be direct and surprisingly blunt. One small example that I experienced: when walking to dinner through the streets of Yokohama, where he now is based, I remarked about how funny it was from an American perspective that Japanese pedestrians will never cross the street against a red light—even if no cars are in sight. They will patiently wait for the light to change. That has never changed in the decades I have been visiting Japan. In that same situation, Americans, and particularly New Yorkers, would charge across the street even if the light were still red against them. What did he think about that? "We respect the rules," he said. "You don't." He thus crystallized a complex cultural riddle in a few words. That's very unusual for a Japanese, many of whom are uncomfortable speaking too candidly lest they hurt someone's feelings.

On the job, Naitoh is like the maestro of an orchestra. Although an engineer, he does not possess deep vertical knowledge of all the different technologies that go into computers. His great skill has been in building teams of people who do possess those skills and then integrating and organizing their efforts to create tangible products, often reaching across geographic and cultural lines to get the job done and get it done on time.

I went to Japan twice to meet Naitoh and spent many days and evenings with him. He is the very antithesis of Steve Jobs or Michael Dell or Steve Ballmer, or more recently, Mark Zuckerberg or Elon Musk. There is little showmanship and no flashy lifestyle. He has a forceful personality, but his brand of innovation is very careful and step-by-step. Acting very differently than some of the tech world's rock stars, he tries to make quick personal connections with the people he meets. He listens to them as they try to solve complicated problems. He builds consensus from the ground up, not by imposing orders from on high.

It's not too much of an exaggeration to argue that there are geopolitical implications to what Naitoh has accomplished. Partly because of his role, the ThinkPad emerged as a unique example of US-Japanese cooperation even during times of great tension in the relationship between the two governments. IBM's personal computer division owned and operated the Yamato Development Laboratory where Naitoh was mostly based, and where the ThinkPad was first developed. There are no other comparable examples of Japanese development labs creating major products for American parent companies. Japanese consumer electronics companies—Toshiba, Sony, and Matsushita (Panasonic)—competed against their American counterparts. Although those companies retain strong positions in semiconductors, sensors, batteries, and screens—the guts of any device—they have largely lost the battle to dominate world markets with their brand names. They became isolated and lost touch with the end user. The ThinkPad, in contrast, still remains an iconic name around the world, at least in part because of sustained collaboration between Americans and Japanese.

When IBM sold its PC division to Lenovo, Naitoh played a role in maintaining the morale of his engineers and keeping the Yamato lab intact. There were huge uncertainties over what management—based in Beijing and North Carolina, where the IBM PC Division had been headquartered—would do with the ThinkPad team. Would they be simply closed down? Would all the functions be moved to China? Would the Chinese drive down the quality of the ThinkPad, which the Japanese engineers considered almost sacred? Naitoh helped navigate that transition from US ownership to a company of Chinese heritage, partly by developing a relationship with Lenovo CEO Yang Yuanqing.

It worked. Lenovo embraced the ThinkPad and has invested in it and expanded its sales beyond anything that IBM ever did. IBM had a fundamentally ambivalent relationship with the ThinkPad and the personal computer division of which it was a part. IBM's top management was oriented toward mainframe computers that rarely broke. But personal computers including the ThinkPad suffered from early quality problems. The PC division was like an unruly stepchild that refused to play by the parent's rules. Under IBM's ownership, only twenty-five million ThinkPads were sold. But under Lenovo's control, that number increased by seventy-five million. The ThinkPad is once again a unique example—this time—of collaboration among Japanese, American, and Chinese cultures—a rare feat in the often-difficult relations among the three governments and business sectors. The United States, China, and Japan are now the world's three largest economies, in that order, so understanding the interactions among them is crucial for anyone who wishes to have a sense of what is happening in the global economy. It is no longer a Eurocentric planet.

This book was researched and written exclusively on Naitoh's and my X1 Carbon ThinkPads. We used the machines exactly as they were intended—collaborating over great distances and thirteen time zones to create a fifty-thousand-word document. For me, the notebook enhanced my personal productivity mostly because the keyboard is a writer's keyboard. When I conducted interviews, either in person or on the telephone, my fingers could move faster than on other keyboards because of how the keys have been optimized. I was able to capture what people told me without having to tape record and transcribe the conversations—a painful, time-consuming exercise. The machine's ability to boot up in seconds also was valuable. That may seem trite, but it was not. It greatly facilitated my engagement because I was not worried about having to sit idly, staring into the screen, while the ThinkPad went through a laborious start-up sequence. And for the first time in my career, I made extensive use of the machine's ability to back up my documents in "the cloud," the world where large server farms store enormous amounts of data. In writing this book with Naitoh, I became convinced that the ThinkPad is the finest platform for any writer doing serious work. It was my distinct honor to work with the man who played the leading role in creating it.

Bill Holstein,
Cortlandt Manor, New York

Chapter 1:
How My Fascination Began—
and How It Saved the
Big Boss

I've always said that my personal default mode came from my father. By that, I mean that he shaped my life so thoroughly that I came to approach problem solving the same way he did. He had a low center of gravity emotionally and was highly analytical, as evidenced by the way he talked about his experiences in World War II. He served in the Japanese Army in China as a mechanical engineer who worked on maintaining Japanese aircraft.

Toward the end of the war, the Soviets swept into Japanese-occupied parts of China and seized territory and men. My father was captured and put on a very crowded train to travel to a prisoner-of-war camp near Moscow. I remember him telling me that the young guards who had to ride on top of the train in bitterly cold weather were worse off than the Japanese soldiers crowded into the cars below them.

I also recall him telling me that the Russians were good people; it was their government that was the problem. Somehow he fell in love with Russian cuisine—he often sought out Russian restaurants in Nagoya in central Japan, where I was born in 1952. Far from looking at his two years of captivity with outrage or hatred, he found a way to make a balanced judgment and to extract something positive from it.

He wanted to work as an engineer, but Japan had been devastated by the war and there were no jobs for him in engineering by the time he got back. He got a job at a newspaper company, and that kept him busy. But he tried to spend leisure time with me, making radios or building lightweight toy airplanes that would fly. This was before the era of transistors, so we had to use old-fashioned vacuum tubes for the radios. I had two sisters and all my cousins were also girls, so his ambition to create a new generation of engineers, rightly or wrongly, fell on me, the only male of the next generation.

I was a bit of what you would today call a nerd. I spent hours whenever I could looking up into the sky at the stars. I was fascinated by them. I joined the junior high school astronomy club, but officials at the school would not allow us to enter the school building at night to look at the stars when they were clearly visible. That struck me as pretty stupid, so I had to do it on my own. I was fascinated by the American launch of the Apollo spacecraft that landed on the moon in 1969, just before I had finished high school. I thought that was truly amazing.

I also became familiar with the first primitive computers. I attended an exposition in Osaka where IBM displayed the prototype of a big printer that could print the complex Chinese characters that appear in the written Japanese language. IBM seemed at the forefront of trying to get that done. For me, computers equaled IBM. They were the same thing.

My father never told me he wanted me to become an engineer. He didn't have to. It was where my passion lay. I was able to win admission to one of Japan's top universities, Keio, in the Department of Instrumentation Engineering.

It seemed only logical that I would join IBM when I graduated. IBM Japan was run by Japanese executives and was seen as a Japanese company. I was thrilled to join IBM's newly established Japanese research center in Fujisawa in 1974. The lab mainly developed terminals for use in industries such as banking.

At the time, Japan was somewhat isolated from the rest of the computing world because of the way our language was written. Let me explain the problem because I think it will give you deep insight into the kind of people we Japanese are. We have an alphabet of fifty-two relatively simple characters called *katakana* that we have traditionally used to absorb Western concepts. We have a second alphabet called *hiragana*, which consists of characters from the Chinese language that we greatly simplified centuries ago. And we also use contemporary Chinese characters called *kanji*. What our written language thus reveals is that we have absorbed many ideas and technologies from both East and West and modified them or improved upon them.

As a result of the complex characters that might require, say, twelve strokes, compared to just three or four for an English letter, our full written language could not be fully computerized. Computers at that time used only a single byte (or eight bits) to express a character. That allowed the computer to represent 256 kinds of characters. It was enough for English and other European languages, but insufficient for the far greater number of kanji characters we use for Chinese and Japanese. So we had to change the computer to be able to handle two bytes of code for kanji characters. The memory capacity was also very small—64 kilobytes—laughable by today's standards and not enough to store font images of a few thousand kanji characters. (People today are accustomed to hearing how many gigabytes of memory their devices possess. A gigabyte is roughly one billion times the size of a byte. That helps show what an incredible explosion in computing power I have seen in my lifetime.)

To attack the problem of computerizing the Japanese language, IBM Japan wanted to develop a two-byte (or sixteen-bit) system, and they created a new product development division to do the job. They asked for volunteers. I raised my hand. I was hungry for the challenge.

I started to analyze the problem. Terminals in those days were dumb, meaning they had little or no intelligence residing in them. They were connected by wires to what we called the controller, which is where the intelligence resided. I started taking apart the controller and looking into the logic circuit, or the brains that controlled it. I realized that the problem was in how printing fonts were stored in the controller. Once I adapted those fonts, the complex characters in our language could be displayed. It was a simple solution. But no one had ever thought of it. Japan could now be connected to the world. Communication terminal technology became my first specialty.

The Advent of Desktop Computing

When the IBM Personal Computer PC-AT first appeared in 1981, it was a seminal moment. A company dedicated to large mainframe computers, or "Big Iron" as it was called, appeared to be acknowledging that a new era had arrived—the age of more distributed computing. It seems amazing to think, but it's true, that for many years, the only way for an individual to obtain computing power was to go to a computer center at a university, for example, and ask the attendants to run your program. You would hand them your stack of cards that had been punched with holes representing your program and the problem you were trying to solve. You might come back the next day to get your results.

The first IBM PC was a great product, but it was not powerful enough to handle the complex characters in our language, which meant we still needed a different version of the PC for the Japanese market. We came out with that unique-to-Japan device two years later in 1983. It was called the IBM 5550. We made it independently at IBM Japan's lab in Fujisawa, which was later relocated to Yamato, a small town southwest of Tokyo.

As time passed and new technologies continued to become available at lower prices, it became clear that IBM should improve the PC's memory and screen resolution. So very gradually, the gears inside the company started to grind away at the challenge of creating a family of new personal computers that was going to be called the PS/2, standing for Personal System 2.

The primary responsibility fell to the company's lab in Boca Raton, which was in charge of all PC development for the world. My job was to redesign the 5550 to be as similar to the PS/2 family as possible. We wanted minimum modifications from the PS/2 with maximum commonality of parts between the two. We tried to make our computers—intended for different national markets—as similar as possible. That made them easier and cheaper to produce and allowed us to introduce new technologies faster.

Nick Donofrio's Crisis

I started visiting the Boca Raton lab in 1985 to get a sense of what the new products would look like. Not surprisingly, the people in Boca Raton were focused on their primary task and did not exactly celebrate me when I came. I was a nuisance because I was asking about things that had no bearing on their ability to produce the PS/2 series. I wanted detailed drawings of the parts they were considering using. They seemed to think I was an odd man from Japan who didn't speak enough English. Frankly, they did not treat me very well. I understood why.

But I learned enough that we were able to get the job done. I was able to converge the two systems, and the 5550 was no longer needed, so it was dropped. That was a clear win. As a result, I guess my bosses saw potential in me. They asked me to transfer to the United States for purposes of grooming me for bigger jobs, a common practice throughout IBM. In April 1987, I relocated with my wife and two children to Westchester County, New York, just north of New York City. I was assigned to the IBM PC Division headquarters on North Broadway in White Plains. I was an assistant to Lou Bifano, who in turn reported to Nick Donofrio, who was in charge of development for the PC division and therefore in charge of Boca Raton.

Donofrio was already something of a legend within the company for getting things done. I was very intimidated to be working in the United States because my English was so poor. When the phone rang, it was my job to answer it. But I found the prospect of answering the phone and trying to communicate in English so intimidating that I often let the phone ring three or four times before picking it up. I hoped that whoever was calling would grow impatient and simply hang up. I had studied English for six years in school and performed well on tests, but like many Japanese, I found it hard to engage in rapid-fire conversation in English.

While I pursued my duties in New York, a different team in Yamato got an assignment to develop another PS/2 product, the P70. This was going to be IBM's first PC for worldwide markets developed outside of the United States. It would become IBM's first "luggable." It was called that because it weighed roughly twenty pounds and had to be lugged around. It was not easily portable.

But the Yamato team ran into a real problem with electromagnetic interference, or EMI. The electronic circuits of computers generate electrical currents and voltages that cause EMI, a harmful emission that also created problems for early television sets. Every electronic device sold anywhere in the world has to comply with the standards set by national governments. In the case of the United States, the key agency was the Federal Communications Commission (FCC).

It was not unusual for an EMI problem to emerge just as a company was getting ready to ship a new product. The reason the problem often showed up late in the game was that we would be testing for EMI at different stages of product development but, as we continued to tweak the design and components, we might inadvertently create new EMI issues. Of course, if the problem could not be resolved, the product could not be shipped. Big financial losses could quickly develop.

My American bosses were obviously concerned about the Yamato lab's problems with the P70 and started asking me to find out what the latest updates were. Officially, I was not assigned to the Yamato lab or the P70 project, but both Bifano and Donofrio wanted immediate answers. Because I could speak Japanese, they turned to me, and I became drawn in. I had to communicate with the Yamato team by phone or fax. Coordinating development of anything on a global scale was twice as difficult as it might be today. But I found that no matter what time of day or night that I called the Yamato team, even 1:00 or 2:00 a.m. their time, someone was there. I believed in working hard, to be sure, but I had never seen a group of Japanese engineers work for as many hours a day as they appeared to be. I thought they were a little bit crazy.

The Yamato team ended up bringing their problem to Boca Raton where they hoped they could find answers. Everyone assumed that Boca Raton would be able to fix it because, after all, they were in charge of global PC development.

But a sense of crisis developed in Boca Raton. The Boca Raton EMI lab was the best in IBM's worldwide system, but even its experts were baffled by the new animal called a luggable. They were accustomed to working with desktop PCs that had a separate central processing unit (CPU), also called a PC box, plus an external keyboard, mouse, and monitor. But now, all of those components had been integrated into a single machine in ways they had never seen before.

Manufacturing had been scheduled to begin, but the problems with EMI refused to go away. Donofrio must have realized he had a real problem, and most likely fearing that his own career was on the line, he put big pressure on the development teams to get the problem fixed.

After the Yamato lab's managers realized the severity of the crisis, they called all engineers from IBM Japan who happened to be in the United States on other assignments and told them to fly to Boca Raton to help out. I was no exception.

When I got to Boca Raton this time, I calmly began looking for the root cause of the EMI problem. I think my approach was guided by my father's influence. If you can define the root cause of a problem, then you have a chance of attacking it. In some cases, there might be two root causes, which makes those problems much harder to solve. But the first challenge is to define root causes.

My English was getting better and I was surprised that people started listening to me. I said the key thing we were looking for was repeatability. If you had succeeded in defining the root cause of a problem and you ran a test and it came out positive, you needed to understand what you were doing and what factors influenced the result. You wanted to be able to repeat the experiment and get the same result to make sure you were addressing the heart of the problem.

I started organizing trials in my head and creating a logical process. If you do something to a system, what do you expect the outcome to be? If you get result A, what does that mean? If you get result B, what does that mean? You should know the expected outcome before you test a piece of equipment. Otherwise, you are wasting your time.

I stayed in Boca Raton for a total of two months working on this problem. It finally clicked. Boca Raton was able to resolve the problem and release the P70 on time. The FCC asked IBM to bring the product to their test lab in Washington so that they could make sure it complied with the regulations—and it passed.

I became popular inside the company's circle of people who worked on EMI problems. The head of Boca Raton's development team came to White Plains and asked my boss to send me down to Florida during the Thanksgiving weekend in November 1987 because they were again having EMI problems on another PS/2 machine, the Model 30. It was assumed that, being Japanese, I wouldn't mind working through an American holiday. That wasn't the case—holiday time was important to me and my family, too. But I accepted the assignment, intending to make it up to my family in whatever limited way I could.

The company paid to fly me to Boca Raton, but I paid to take my wife and two kids because they had never been to Florida. I went to work in a dark EMI chamber for several days, but they got to swim at the beach. They even saw a space shuttle take off from Cape Kennedy, which was to the north. To this day, my wife remembers this as one of her most positive experiences at a time when I was not providing her many joys because of my workload.

In the EMI chamber, I once again used my father's instincts and helped get the job done. IBM was able to ship another new product.

Because there were layers of management between Donofrio and me, I wasn't completely aware of how much he appreciated my work until more than a decade later in the year 2001. Peter Hortensius, who was vice president and general manager of the ThinkPad business unit at the time, tells the story of how he approached Donofrio to ask him if he would consider naming me an IBM Fellow. Being named an IBM Fellow is the highest honor the company can bestow. Recipients are given leeway to pursue the research that most fascinates them.

Donofrio had been promoted and was now a senior vice president, which is a bit like being one of the Greek gods on Mount Olympus. The reverence with which senior executives were treated may have been part of IBM's increasing dysfunctionality, but that's the way it was. Donofrio reported directly to the chief executive officer of the company and was basically in charge of the company's technology strategy. Getting the EMI problems fixed must have helped his career.

Hortensius had to personally go see Donofrio to discuss my appointment. "Nick, we'd like to bring Naitoh forward as an IBM Fellow," Hortensius recalls saying. "Would you support that?"

Donofrio responded: "Naitoh saved my ass in 1987. Of course, you can bring him forward. The products didn't work and Naitoh fixed them. He fixed someone else's problems."

Into the Fire

As of March 1988, my plan was to remain in the United States for another six months, until September. I extended my house rental contract and put in for office space in the new IBM headquarters building that was nearing completion in Somers, also in Westchester County, New York.

But a telephone call changed all that. My boss at Yamato said that although he had planned to let me stay a while longer in the United States, he now wanted me to come back immediately. There was no specific mention of what my responsibilities would be. I had no idea. That's not unusual inside Japanese-style organizations. We are expected to be good samurai warriors—to take orders and not question them.

I got out of the rental agreement and packed up everything within three weeks.

En route to Japan, my family and I stopped in Maui for a few days in the sun. We had a lovely time with our daughter, who was about nine years old, and our son, who was three. Little did I know, it would be my last vacation for more than ten years. As the Americans say, I was leaping from the frying pan into the fire.

Chapter 2:
Proof of Concept:
Showdown in North Carolina

As soon as I got back to Japan, I was told that I would be in charge of the development of battery-powered PCs. One major impetus for the decision to push into this product category seemed to be IBM's relationship with Harvard Business School, which was a leader in pushing the adoption of technology in business. It was at the business school, for example, that VisiCalc (for "visible calculator") had been created, which was the first electronic spreadsheet. VisiCalc, in turn, gave way to Lotus Notes, which was considered cutting edge at the time.

In May 1984—still the Dark Ages in the information technology (IT) field—the business school's Professor F. Warren McFarlan wrote an article in the *Harvard Business Review* entitled "Information Technology Changes the Way You Compete." It described a series of cases where IT gave companies a competitive edge in the marketplace by helping them differentiate their products.

Very few people in the business world could see, at that time, how important computers were going to become. McFarlan was thus an early apostle of the idea that computing would revolutionize the world of business.

Starting in 1984, we sold luggables to the business school, which wanted all eight hundred students to have their own computers. They weighed as much as thirty-five pounds, depending on the models and on how many other pieces of equipment were included. The school provided students with something similar to the wheeled pull-carts that airline flight attendants used to use to maneuver their personal luggage. IBM had to establish a repair center at Harvard because the machines broke down so frequently. Their range of use was limited. The moment you unplugged them from the wall, they ceased to function. The screens displayed yellow characters on a black background. Imagine that, children, a world without full color.

These machines relied on MS-DOS, which stood for Microsoft Disk Operated System, and later on black-and-white versions of Microsoft Windows. There were no brightly colored icons to guide a user. The text was not crisp but rather slightly blurry. And they did not support many of the functions we now consider basic, such as saving music or pictures. Millions of young users of tablets and smartphones today cannot possibly imagine these limitations.

Perhaps because our luggables had so many problems, the business school turned to Zenith, the consumer electronics company that is no longer in business today, to supply slightly better computers starting with the 1987 school year. IBM, which had prided itself for its affiliation with Harvard Business School, wanted to reopen the door there and get back in. They soon got their chance when the school decided to request bids from different manufacturers so that the next generation of computers would be available for the start of the fall semester in September 1991.

The dean of Harvard Business School was John McArthur, but the driving force behind the idea of mobile computing seemed to be McFarlan. He did not have any official title such as senior vice president of information technology or anything like that. But because the dean trusted him and senior faculty trusted him, he was in charge of taking the bids and ultimately making a decision.

The new machines were primarily intended to help the students develop case studies, which is the core of the curriculum at the business school. Students could not use the machines in regular classes, but were expected to use them during a rigorous four-hour final exam. During those exams, the students were given case studies and encouraged to use VisiCalc on their computers to analyze data. It normally took two hours of reading and analyzing before the students were ready to start writing their conclusions, also on their computers. There was no access to the Internet because the Net was still too primitive.

For years, these exams had been logistical nightmares because extra power cords had to be provided to power the luggables and technical support had to be present in case a machine crashed. The need for a better alternative was clear.

We didn't know it at the time, but McFarlan's decision boiled down to considering three companies—Apple, Digital Equipment (DEC), and us.

John Sculley from Pepsi had arrived at Apple as chief executive officer and pushed out Steve Jobs in 1985. McFarlan thought Apple had "terrific products" but said later, "We weren't sure the company would survive." DEC was located in Boston, and a key provider of the venture capital that DEC had raised was an investment fund run by a faculty member at the business school. As a result, they had a solid connection into the business school. DEC was making a computer called the Rainbow. But McFarlan did not believe DEC was truly committed to the Rainbow and, in fact, a few months later, it was discontinued.

So we won the bidding. But could we really take the next step to truly portable computing with much smaller form factors than anything we had ever built? We had made the aggressive decision to sell something to Harvard that we didn't know could be made. And could we meet McFarlan's quality specs? "We want to make damn sure whatever we buy actually works," he warned. Frankly, we did not know the answer.

Project Aloha

To take on the Harvard challenge, the company decided to give the Yamato lab a chance to distinguish itself. I would be working with the same crazy engineers I had previously been communicating with from New York. It was a group of about fifty people. We were competing against two other IBM development labs, Boca Raton and Raleigh, and sometimes cooperating with them.

I don't know whether top management at IBM realized it consciously or not, but there was very strong logic behind the decision to let us Japanese engineers take on the challenge of creating portable PCs. Nearly every part of a portable computer would have to be made smaller than it was in a desktop—and making things smaller is a deep part of our culture. Japan is the same size as California but has 125 million people, three times as many as California, and we are crammed into much tighter spaces than almost anywhere else in the world because so much of our land is dominated by very steep volcanic mountains. Flat land is scarce. So virtually every piece of flat land is used for something. That forces us to live in smaller homes and drive smaller cars. Our culture encourages us to make everything lighter and smaller, which is the opposite of the United States, where bigger things are desired. Big people need big trucks to go shopping in big stores with wide aisles.

We also had a very different style of innovation. I call it bottom-up. It's not too far-fetched to argue that it had something to do with our rice farming culture. We had to cooperate at every stage of planting and harvesting, which meant we had to have a degree of consensus about what we were doing. In fact, the word for creating consensus in Japanese, *nemawashi*, literally means "binding the roots of rice plants." Our mind-set comes from the practice of agricultural cooperation.

The style of innovation I saw in American companies (and later in Chinese ones as well) was that a big leader would declare a vision of what he or she wanted to accomplish and announce it to the world. That would start a process of trying to determine whether it really was possible to make the new product and whether the market really would buy it. I call this the individualist hunter's approach, which is definitely top-down. It may work some of the time, but not all the time.

One last thing I should say about culture is that we Japanese embrace the concept of *kaizen*, which is usually translated as "continuous improvement." Toyota Motor made the term famous with its lean production methods. But it wasn't just a Toyota term. The desire to constantly make things better, year in and year out, without ever quitting, is deeply embedded in Japanese culture.

There were two portable notebooks on the market already from Toshiba and Compaq. My job was to make IBM the third member of this club. This was going to be a make-or-break moment for Yamato because we wanted to carve out a role for ourselves in IBM's global operations and not be confined to making products only for Japan. If we couldn't distinguish ourselves, we would be vulnerable to restructuring, or worse.

I officially started on the project, code-named Project Aloha, in late 1989. My deadline for completion of the work was September 1990, roughly a year away. This product was to be called the Laptop 40 SX or L40 SX for short. We didn't know it at the time, but this battle was in some ways a test case to see if a computer of these dimensions could be achieved.

We started with certain advantages. One was that our keyboards were descendants of the old IBM Selectric electronic typewriter, another piece of technology that many young people today have never seen. Keyboards were designed by the keyboard manufacturing division in Lexington, Kentucky, which is where the Selectric used to be made. The keys were concave and, to put it simply, they felt good to anyone who typed a great deal. We took what we learned from Lexington and refined it again and again over the years. We learned to design the keys to provide a sense of feedback to the user. We could diagram how the keys should respond to human touch to provide the most satisfying feeling. If there was too much resistance, the result was a heavy-feeling keyboard. A gentler force feedback curve resulted in a lighter touch.

We also knew that there was an ideal click sensation—the fingers should depress a key until it reaches an ideal point to give way and click. It should not hit the computer beneath the keyboard. That would be unsatisfying and possibly harmful to the health of the fingers. Like a mechanic tunes the engine of a car, we could tune the way the keys responded to the touch. No one has ever been able to duplicate our knowledge of the tactile sensations of a keyboard.

But other technical problems, particularly size and weight, were overwhelming. We had to figure out how to lay out the motherboard to fit within the requested dimensions, which obviously were a fraction of the size of a full PC. The motherboard included the main memory chip, other semiconductors with supporting functions (called the "chipset"), and power supply components. In larger standing computers, finding the space for these components was never a problem. Engineers took as much space as they needed. But now we had to figure out how to minimize the space.

We had other unique requirements to resolve, such as figuring out how to build in a monochrome liquid-crystal display screen, a battery pack, a 2.5-inch hard disk drive, and a slim floppy disk drive. These floppy disk drives don't exist today, but they were very important at this time. With a laughable capacity of just 1.44 megabytes, they represented the way you loaded anything into the computer and removed it. You couldn't email documents at that time.

We in Yamato were responsible for developing the guts of the new machine, but the project involved several other geographies. For example, mechanical design was done at Boca Raton laboratory in Florida; they had overall responsibility for the project. An IBM plant in Research Triangle Park, North Carolina, and one in Greenock in

Scotland would be responsible for manufacturing. Our relationships with these other labs were always a delicate balance. IBM based its business model on creating conflict among different teams of people to see who could create the best result. They had a conscious policy of fostering that conflict to create internal competition. Sometimes it worked. Sometimes it didn't. But that was their philosophy.

A large number of partner companies were also involved in the development stage of the project because we needed technologies from them. Managing all the different time zones was challenging, but managing the different cultures was even more difficult for me because I had spent so little time outside of Japan. We Japanese have a very strong culture and know what to expect from one another. But we are often surprised when Europeans or Americans or others don't understand what we implicitly recognize. I had to work on adopting a different, more Western, style. Some of my American friends within the company would tell me I should not just sit in a meeting until I was called upon, which was my cultural inclination. And rather than trying to resolve conflicts quietly behind the scenes, I learned it was important to let others understand what I knew and what I did not know—very clearly. I had to learn to let them know exactly what I needed from them. I learned to speak up at the right time at the right meeting. In many ways, it was un-Japanese—but it was necessary.

I had to be constantly in motion. Although officially I had returned to Japan, I made a point of being at the location of maximum workload during the various phases from development to production. For instance, at the initial concept-design stage, I would stay in Boca Raton or in California, where a collaborating company, Western Digital, was located. Western Digital would be making the chipsets.

My job was essentially to be a systems integrator, making sure that people in different technological disciplines were working in sync with one another, and to keep everything moving and on schedule. I had some deep technical knowledge but not enough to make all the technologies work together. Once again, that required me to work hard on my communications skills and on understanding what people were really saying.

One of our problems was minimizing interference from unwanted electromagnetic wave radiation, the same problem I had helped the Boca Raton lab conquer. Now the problem was even more severe because we were creating a much smaller machine. (The problem was not fully conquered until the late 1990s when spread-spectrum technology was invented.)

We engaged in lots of trial and error, but I tried to guide the process toward a logical conclusion. We found that the torque with which a single screw in the motherboard was tightened might influence the EMI problem. It was maddening.

We were trying to fit various components, each interfering with the others through wave emissions, into a very small space. If one part changed in size or shape, then other parts also might have to change. Even if we used the same parts, slight differences in how a coil was wound in each part could lead to a completely different result.

We had to develop a problem-tracking system to manage the dozens of issues that would be pending at any one time. When a problem occurred, we assigned that problem a number and entered it in a database. The first stage was to "open" the problem. Then the next stage was for the team of people responsible for the part or the system to declare that they were "working" on the problem. When they had a solution, the party who had identified the problem would "verify" that it had been fixed. Then we "closed" the issue. So we managed all our problems through the steps of Open, Working, Verify, and Close.

Naturally, there were deadlines. The person who discovered a problem set a deadline for fixing it. Based on this, a schedule was drawn up that set a deadline for verification and for when the case was to be closed. The pressure mounted daily. My challenge was to keep track of all this and determine whether we were running into a disaster or whether we were on track. Quite honestly, it was often difficult to tell.

To get the job done, I put enormous pressure on my engineers and they put enormous pressure on one another. I became famous, or infamous, for telling everyone, "No excuses." I also applied enormous pressure to myself. As they say in the movies, failure was not an option.

The Personal Struggles

Because of the constant demands, everyone worked incredible hours during this period. I realized how much pressure I personally was under when late one night, after eleven, the taxi that was taking me home was nearly hit by another car. The thought actually occurred to me that it wouldn't be so bad to have an accident and go to the hospital. At least there, I could have rested.

I was aware during this period of time that the burden of caring for my children shifted almost entirely to my wife. When a problem came up with one of the children, or both, she had to handle it. I usually wasn't there and might have been in a different time zone with only a few minutes to talk on the phone. If I was out of the country, IBM would pay for only three minutes of an international phone call per week. Those calls were much more expensive back then.

Of course, I wasn't the only one struggling. During one period, Kenshin "Ken" Yonemochi (who was my No. 2 at the lab and who went on to become the worldwide director of product assurance for ThinkPad development) was managing design activities at the Yamato lab on a de facto basis when I was away. Though a keyboard specialist, he had general oversight of mechanical design. As I returned to Japan, I learned he had been hospitalized.

I rushed to visit him because sometimes Japanese executives or engineers literally work themselves to death, a phenomenon called *karoshi.* I was relieved to find out that his condition was not serious, but he was hospitalized for a week with severe headaches. He had been working every day from 9:00 a.m. to 2:00 a.m., or seventeen hours. He was taking only one personal day every two weeks.

I am often asked why we worked so hard. What in particular drove me? Was it a desire to help Japan recover from war and revive its economy? I certainly was aware of the need to improve Japan's economy, and I was aware of Ezra Vogel's book, *Japan as Number One,* which appeared in 1979. That book was generally recognized as representing the first recognition in the Western world that Japan was making rapid economic strides. But I don't think that Japan's overall economy was the main motivator. Nor was the motivation based on the desire to get rich, which is what drives many entrepreneurs in Silicon Valley. I wasn't getting any stock options or anything like that.

The simple truth was that I was fascinated by the possibilities of new technology. And I had developed a fierce loyalty to everyone at Yamato. We were like a family. Together we wanted to prove something to ourselves, to IBM, and to the world.

The Drive to Produce

Everything back then was a first: We developed our own power management algorithm in the basic input/output system (BIOS), which is invisible to users but controls the computer's peripheral devices such as rechargers and external disk drives. That algorithm work was very different from other types of work we were doing because it involved software, not hardware. We defined a benchmark to test the life of batteries, which hadn't previously existed. Industry specifications and components were available for desktop PCs, but not for notebooks. We really did take pioneering steps in defining so many of these features and standards, which today are taken for granted.

One of the reasons we were able to do all this was that we were located close to subcontractors in Japan that made many parts, and we could consult with them intensively. We also had a trial production line in nearby Fujisawa and could experiment to see what actually could be manufactured. Some experts call such setups "feedback loops." Because information could flow back and forth easily, we could constantly refine and innovate. It's almost impossible for someone sitting in a lab by himself or herself to truly innovate without knowing how all parts are manufactured and without knowing whether a particular design can actually be manufactured on an assembly line. You can have a great idea in the lab but if an operator on the assembly line can't physically make the part, your idea is not worth very much.

We could go to Fujisawa to watch an operator actually build something based on our designs. We, as engineers, might think it was easy to make a certain subassembly, but the parts were often very small and human fingers found it difficult to handle them. That created assembly error and obviously affected quality. We could get this feedback from the test line very quickly, much quicker than if the test manufacturing had been located in China, for example.

One of the mistakes that some American computer companies would later make was outsourcing the design of their products to middlemen in Taiwan and also outsourcing the manufacturing of the products to third parties in China or elsewhere. One of the senior executives at Lenovo, Gerry Smith, would later say that those companies "outsourced their brains," which is why they missed important shifts in the market such as the move to tablets or smartphones. They could not respond fast enough to the shifting tastes of their customers.

Showdown

We had not yet brought all the parts together in Japan to make a completed product, but finally, we thought we were ready to start mass production in North Carolina as scheduled in the fall of 1990. It was time for me to station myself at the North Carolina plant, which had formerly made display monitors for IBM. I fully expected that we would be shipping portable laptop computers to Harvard as of September, as scheduled. I was there to integrate everybody's efforts in all aspects of hardware and software.

But I soon discovered that, despite all the preliminary work we had done in Japan, our trial production runs in North Carolina would suddenly stop. The parts were not fitting together well enough to actually manufacture a product. It's obviously a very bad thing when a production line just stops. That forced us to make changes in the specifications of the product and the parts. If something can't be manufactured, it has to be redesigned. That took precious time.

Another set of issues revolved around the fact that this was the first time that we were working with the chipsets from Western Digital and we needed precise coordination with them. We at the Yamato lab had gotten to know Justin "Jud" McCarthy when he was posted to Yamato for two years. He was about fifteen years older than me, and I had come to think of him as a mentor. Partly because we used to drink beer with him on Friday nights and listen to his colorful stories, we trusted him and so we asked him to relocate to southern California to help us coordinate with Western Digital. He was the senior IBM man on the ground there. It took many iterations back and forth, between Jud in California and me in Raleigh, to work out problems in how the chipsets worked with the rest of the machines we were trying to build.

Moreover, this was the first time the production line operators at the North Carolina plant were making products with tolerances of less than a millimeter, which is 5/127 of an inch. A tolerance is how much space you allow to exist between parts. Larger tolerances make things easier to make.

But because we had to cram so many components and parts into such a small space, we had to use very tight tolerances. We might have been able to achieve that more quickly if we had had more experienced workers, but we had many new employees working on the line. So we had to deal not only with design problems, but also with inexperienced operators.

The visual representation of failure was the red tags that inspectors would place on machines they had rejected. Sometimes it was operator error; other times the parts just didn't fit right. The red tags were awful to see.

Whenever I saw those tags, I tried to analyze what had gone wrong. Was it a problem of how the parts fit together or was it a mistake made by an assembler? When I found a very common operator error in a machine, I had to go to the line and say, "You should do this; you shouldn't do that." Once I started this process, there was no way I could quit. I was filling a vacuum that had to be filled.

I even tried to conduct tests on how to fix problems, such as the leaking of energy from the coil (the power supply), which created errors on the magnetic head in the floppy disk drive. With my red Swiss Army knife, I cut a piece of aluminum foil to try to block that energy leak. It didn't work, but that's how desperate I was.

The days and weeks dragged on. This was one of the most difficult periods of my life, and it took a huge toll on my team as well. On the day before New Year's Eve, I discovered a problem and wanted a quick fix. I remember calling Ken Yonemochi in Japan from Raleigh late that day, which would have been early morning on the day of New Year's Eve because of the time difference, to let him know that I had sent a fax about some problem. Of course, he would have to go in to the office to read that fax. Essentially, I was asking him to go to work on New Year's Eve, an important day for families in Japan. We don't celebrate Christmas or Thanksgiving, so there are relatively few times that Japanese families can count on being together. On top of that, my fax said that the problem had to be taken care of by the next day. Yonemochi called everyone in, and the Japanese development team worked straight through the holiday.

I was very sorry to force my team to work through the holidays, but we were under intense pressure as we entered 1991. I felt that I was fighting for the Yamato lab. This could be the moment when we carved out a unique role within IBM and escaped the shadow of Boca Raton and the other development lab in Raleigh.

About this time, I received a call from Jud McCarthy on the West Coast.

"How are you doing, Naitoh-san?" he asked.

I knew by then that when Americans ask this question, they don't really want a complete explanation. They want to hear a quick "fine" or "great" or something very brief. But I decided to tell McCarthy the truth.

"I think I'm dying," I told him, matter-of-factly.

"Well, look out the window at all those pine trees in North Carolina," he replied. "You won't have to worry about finding enough wood for a coffin."

I knew he was trying to be funny and cheer me up. I was able to manage a small chuckle, but that was all.

In January, we tried to start manufacturing again. The production line was operating in three shifts around the clock, and line operators were able to clock out when their shifts ended. But I had to greet the people of the next shift to tell them what we had learned on the previous shift. I didn't have time to return to my hotel, the Marriott in Research Triangle Park in Durham. So I found a conference room on the mezzanine level that overlooked the manufacturing line through a glass wall. The conference room had been used to show VIPs the line-making display terminals.

Mark Cohen, who was there on behalf of the Boca Raton lab, describes how one night he found me asleep on the floor with my head resting on my open briefcase. I must have opened it to allow my head to rest on papers that were softer than the surface of the briefcase. Eventually someone gave me a sleeping bag and a couch appeared, so I could catch a few hours of sleep at a time. Some of my colleagues thought I never slept at all, but I had a secret place. I would have to shower every two or three days, so I would rush back to my room at the Marriott, but that was all I used it for. It was during these rare breaks that I would turn on my radio and try to catch any news from Japan. This went on for about two months, seven days a week. Telephone conversations with my wife back home were very brief. There was no such thing as email or Skype.

Cohen and others from Boca Raton were trying to keep Harvard Business School happy. They rented a plane and flew from Raleigh to Logan Airport in Boston on a regular basis with a handful of machines we thought would work. We wanted McFarlan and others to see the progress we were making. We put some of the machines into the hands of students and transferred the data from their old machines to the new ones, but then some of the new machines would break down. It was a mess.

We kept working on the problems, one by one. Finally, just as the different sections of an orchestra finally come together to make beautiful music, we finally received authorization to ship on March 13, 1991. This meant the L40 SX had received final approval from all levels and departments within IBM, particularly the product assurance people, and we could start shipping to Harvard in quantity. By coincidence, this happened to occur on my birthday. Knowing that, my Raleigh coworkers broke out singing "Happy Birthday."

They then took out some items and gave them to me, saying they were "presents." In fact, they were parts that had failed during development! Everyone had set them aside just for this occasion. One item, for example, was a cover that had blocked the slot of the floppy disk drive. I was so exhausted that my sense of humor was not so good. I begged people to not give me any more parts that hadn't worked. I couldn't bear to be reminded.

We had prevailed. We had created IBM's first battery-operated notebook computer—and we had learned how to actually manufacture it—just by the skin of our teeth. The era of luggables was ending.

"The L40 SX barely met the specs, but it did meet them," McFarlan would recall later. It was lighter, smaller, and had more battery power than its predecessors. Students would no longer need their stewardess pull carts. As McFarlan put it, "IBM was back in the game."

Chapter 3:
Project Nectarine: The ThinkPad Survives a Ghostly Encounter

We thought we had solved the problems of building a battery-operated notebook, but IBM did not make a big sales, branding, or marketing push behind the L40 SX. I felt as if top executives at IBM saw what we had achieved and decided it was time to quickly create a new version that contained more of IBM's own technologies, rather than merely assembling other companies' technologies. Some critics inside the company and in the design community also said they thought the L40 SX was not attractive in terms of appearance. Truthfully, appearance had not been my top priority. I just wanted to prove that it could be engineered and built.

Because of the design issues, this next iteration was going to involve the famous German designer Richard Sapper, who was based in Milan, Italy. Top management, hungry for a victory in the marketplace at a time when the company had started losing money, also seemed to decide that they would aggressively promote the new product.

It seemed that executives at IBM Japan were able to cite my role in building the L40 SX in North Carolina as grounds for allowing the Yamato lab to take the leading role in developing this very important new product. I could not see what was taking place several management layers above my head, but the infighting and maneuvering among the senior vice presidents and division presidents must have been intense.

This notebook was going to be called the 700C and it was going to be lighter and somewhat smaller than the L40 SX, with more powerful processing power and longer battery life. Having gone through the hell of creating the L40 SX, I felt confident in my team and myself—maybe a little too confident.

The project's code name was Nectarine. The name of the fruit had nothing to do with the actual project, but we loved having code names. We would now be able to use a 10.4-inch color thin-film-transistor liquid-crystal display (TFT LCD). These displays were being made at a 50-50 percent joint venture between us and the Japanese company Toshiba called Display Technology, Inc. The C in 700C stood for "color." We didn't know what percentage of the 700Cs would have color displays and how the market would receive them.

Nowadays, color LCDs are commonplace, but at the time they were amazing to behold—and considered very expensive. We know now in retrospect that the advent of color changed the way people felt about engaging with their devices. It added an element of emotional intensity. But at the time, amazingly, we were not sure that color would ever be accepted in the market. We thought it might be too expensive and customers would not be willing to pay for it. (Over time, of course, the cost dropped considerably, and the marketplace reaction was overwhelmingly positive.)

We worked closely with Intel to manufacture our version of their new 486 semiconductor. Our version was code-named Bimini, and it represented a major advance over the previous generation of Intel's 386 chip. Moore's Law was very much a reality—Intel cofounder Gordon Moore had predicted as early as 1970 that the power of semiconductors would double every eighteen months to two years—and we were right in the middle of that revolution.

So we were going to design the system on the basis of IBM's own semiconductor design and therefore first had to develop our own chipset. This was the first time we had ever tried to develop an entire chipset for a notebook. The reason we needed chipsets was that the main memory and processing semiconductors, at this point in history, were not integrated with all the other functions such as the display controller and the memory controller that were necessary to run a battery-powered computer. So we had to surround the main chip with a set of other chips that performed these specialized functions.

To save time and money, we started with a chipset code-named Spice that we had developed for other products in the Japanese market. Mitsuo Tabo, the technical leader of chipset development in those days, reminds me that this new chipset became nicknamed Spice II because it evolved from the first Spice. We thought we were being very clever by giving each chip a spice code name. The display controller chip was Cinnamon, the clock in the central processing unit was Mint, the memory controller was Laurel, the peripheral controller was Allspice-II, the low-power small computer system interface controller was Basil, and the power management and system management controller was Ginger. IBM Japan made four of those chips, but another company, Seiko-Epson in Japan, made Mint and Allspice on the basis of our designs. All this represented a big advance over simply buying chipsets from Western Digital.

However clever we were with the names, it was the first time we were building such a chipset for a notebook and, in retrospect, we should have anticipated problems. (In later years, after IBM sold its semiconductor business, we relied entirely on Intel. As their processors became more sophisticated and "integrated," they gradually absorbed the functions that chipsets performed.)

Another major decision we had to make related to color. Up until this point, the vast majority of computers had been white or beige in color. But Richard Sapper was determined to make this one black. He and members of the IBM Corporate Industrial Design team visited me and announced that the machine would be black. Sapper's style

45

was to design things that looked very simple on the outside, or at first glance, and he preferred black to emphasize that simplicity. But then as someone started to engage with the item he had designed, they would be surprised by what designers call its articulations—how many different and complex things the object could do or offer. For emphasis, he liked the points of engagement to be red in color. As a result, many of the objects he designed were black with red accents.

At the same time, we made a major decision to include the Track-Point, which was developed by Ted Selker at IBM's Almaden lab in the United States, in the middle of the keyboard between the G, H, and B keys. The TrackPoint was controversial to people who preferred to use a mouse to guide the cursor, but we wanted the user to be able to control the keyboard with shorter, more efficient movements than a mouse required, and we wanted it to be possible for our laptop to be used on the meal tray in an airplane. That was a tight space. We figured that if someone had to use a mouse on a pad beside the computer, there wouldn't be enough room. The first generation of Think-Pads had only the TrackPoint, and no touchpads. Those didn't arrive until after 2000.

An enormous amount of work went into figuring out the right type of material and design for the TrackPoint, so that the human finger could easily manipulate it. One problem was that human sweat could make some materials slippery, so we experimented with tiny extrusions that created a more textured surface, as you can see in the photo of the TrackPoint on the cover of this book.

Sapper was determined that the TrackPoint would be red, and that started a fight with the corporate ID people. David Hill, who became head of ThinkPad design in 1995, tells the story of how the corporate ID mavens told Sapper the TrackPoint could not be red. Their improbable reason was that the color red was reserved for emergency shutdown switches on the backs of mainframes located in data centers. They didn't want to confuse customers.

"You've got to be kidding me," Sapper responded.

To get around these naysayers, Sapper created a TrackPoint that was magenta in color and gave it an official product color code. Let's say it was TP333. That satisfied the corporate ID department; however, in each iteration of the TrackPoint, as it kept getting refined, he would make the TrackPoint a little bit redder. When the corporate ID people challenged him, he would respond that it was the correct color because it had the right code, TP333. It clearly was red—but he had cleverly outmaneuvered them.

The red TrackPoint soon became one of the most visible design features of the machine. Sapper's design of a black ThinkPad, with a red TrackPoint in the middle of the keyboard, would become a visual icon around the world. The design has often been compared to that of Porsche automobiles, which I adore.

The ThinkPad name itself was not decided upon until just prior to the sales launch. The origin of the name was derived from the motto "Think!" that had been introduced as a slogan by IBM Chief Executive Officer Thomas J. Watson Sr. in the 1920s. New IBM employees were given paper notepads with the word "Think" written on their cover.

Another piece of the equation that fell into place late in the process was that Toshiba came to us and said they could not buy half of the 10.4-inch color displays we were making together in our joint venture, which had been our agreement. They obviously did not think they could sell them. We were not sure either but we agreed to take 100 percent of the color display production. It turned out to be a great bet. It meant that we could make more 700Cs, and it meant that we were the only company to offer color screens of that size for a period of time. It turned out that these screens were one of the technologies that attracted the most attention from customers. We just got lucky to be able to increase our supply of them.

It appeared as if everything was poised for a great leap forward in terms of color, power, weight, and battery life. Many different parts of IBM were contributing toward what we were trying to achieve. It felt very different than the near chaos I encountered while building the L40 SX. Rather than being isolated and ignored, we were now at the center of the company's attention. The product would be made in Fujisawa, Greenock (Scotland), and North Carolina. Thanks to the hard lessons we had learned with the L40 SX, manufacturing would not be a serious challenge.

The Last-Minute Ghostly Encounter

We had finished a complete prototype of the 700C and felt that we were coasting downhill to victory. We were in the final testing phase in early 1992. The make-or-break Go Ahead date to proceed with production, which we called the GA, was about three months away.

But one day, we noticed that the machine started getting hung up and stopped operating. It would freeze. This occurred when we kept repeating a screen animation program over and over to put stress on the system. The freezing didn't seem to happen with any regularity, which made it harder for us to identify the problem. A regular pattern would have suggested answers, but this appeared to occur randomly and it only happened once over the course of two to three days. It was like there was a ghost in the machine.

We started to use our data scope, which can record the level of different signals produced by the chipset. The scope had cupper lines that we attached to the chipset. From experience, we knew that different patterns of signals meant that different specific events may have occurred. But the scope had limitations. It had limited memory, which meant we could not monitor the pattern of signals for very long. Another limitation was that the scope could "hear" the pattern of interaction among chips, but it could not detect what was going on inside the silicon of an individual chip. There were three primary teams engaged in trying to understand what was happening—a software team, a chipset design team, and a group of electrical engineers.

We couldn't detect anything wrong with the hardware, so I assumed we had a software problem. The members of my software team and I went into crisis mode trying to understand and fix the problem. We got very little sleep.

By now, I had learned how to operate on little or no sleep. There were nights when I caught an hour or two of sleep in the office, but as a matter of routine during this general period of time, I would go home almost every night after eleven. If I was taking the train, my wife would pick me up at the station and drive me home, never once complaining. I could drive my own car to work only two days a week because we had limited parking space on the grounds of the Yamato lab, which was located in a residential area in a small town. My other alternative, if I had missed the last train, was to take a taxi home.

Whenever and however I got home, I ate some dinner, took a shower, and jumped into bed. The children were already sleeping, of course. I normally woke up at 7:30 a.m. and had a few minutes with my children, but very soon it would be time for us all to leave for school and work. My wife would drive me back to the station on days I took the train.

My daughter was about fourteen years old and my son was seven in 1992. When I later finished working in this incredibly intense mode for ten years, our children had already grown up. My wife told me in later years that it had been hard to handle a few critical issues the children faced in school without my assistance. She says now, without bitterness, that our children grew up without me. My mother complained more forcefully than my wife because I did not often visit her or write. Today, I am sorry for the impact my work style had on my loved ones, and I realize today that everyone who worked for me paid a similar emotional price.

I do not think, however, that these were bad days. Yes, we worked hard and didn't make very much money. But somehow, these were shining days in our lives. We had big projects! Big challenges! We attacked them as parts of teams and we kept improving. We were winning. I think that's what drove us. It was not a totally bad time in our lives.

The weeks were ticking by as my software team and I hunted for the ghost inside the 700C's chipset. We went through one week, then two weeks, of all-nighters. The chipset design team, who was responsible for the hardware, not the software, was aware of our problem and analyzed test results, but had not been staying up all night.

Suddenly after more than two weeks of delays, the chipset design team found a physical problem in the Allspice chip. We had made a design error. Now I started staying up all night with the chipset design team instead—the software team could get some rest.

At first glance, the fact that it was a hardware problem was bad news. If it had been a software problem, we could have rewritten a few lines of code and been done with it. But a hardware problem meant that the manufacturing process for the chip would have to be changed—and that might require too much precious time if a vendor were making the chip. That was the situation in this case—Seiko-Epson was making the chip based on our design.

But fortunately, Seiko-Epson had what they called a Quick Turnaround Time manufacturing line, and they were able to fix the problem quickly. We got back on schedule. We made the ghost go away. Once again, there was logic in having deep relationships with vendors located nearby.

The Launch

Everything came together smoothly after that near disaster. The ThinkPad 700C had four megabytes of standard memory and was able to operate the 3.1 version of Microsoft's Windows operating system in color. Its battery was nickel metal hydride, which was very similar to what is used in automobiles. It was primitive when compared to today's batteries but allowed the user to operate the machine for three hours and twenty minutes after it was fully charged, which was considered a lot at the time. The hard disk drive slot was located on the front right bottom of the machine, so that it could be removed or inserted while the user was seated on an airplane. If it had been on the side of the machine, that would have been much too awkward.

IBM launched the ThinkPad 700C to the industry as a developer's release in April 1992. The company did some very clever things during this period to build excitement around the ThinkPad, including sending one notebook to archeologists excavating an ancient Egyptian city called Leontopolis that summer. It was incredibly hot and dusty, but the archeologists reported that the ThinkPad, was "rugged enough to be used without special care in the worst conditions Egypt has to offer."

For the general public, IBM unveiled the 700C at a press conference at its Manhattan headquarters building in October 1992, just eighteen months after we had perfected the L40 SX.

I was not asked to attend the launch, and that was fine with me. There were many levels of executives above me and it was their job to engage with the media. Japanese culture does not encourage us to stand up and brag about what we have achieved. Other people are supposed to recognize what you have done. Let the results speak for themselves. I knew what I had done, and my colleagues knew what I had done. That was all I wanted.

At that time, no one knew the extent to which notebooks would be embraced by the market, or how much market share they could capture. Al Gore, then vice president of the United States, talked about the "great information superhighway." But the Internet had not yet really taken off.

IBM was losing money again in 1992 and people were being "restructured" out of jobs, which we Japanese find particularly shocking because traditionally we have believed that our companies have a responsibility to offer lifetime employment. In December, IBM sold its hard disk drive business to Hitachi, which appeared to mark a retreat from manufacturing. By the end of 1992, the number of employees had fallen to 301,542, as compared to 344,396 from the previous year, according to IBM's archives. The mother ship seemed to be sinking, but the ThinkPad was like a bright light that we could all follow.

Writing with the benefit of hindsight on the twentieth anniversary of the ThinkPad, here is how David Pierce of *The Verge* described what happened when IBM presented the 700C at what was then called the COMDEX tradeshow in Las Vegas in November 1992:

> The 700C quickly became the belle of the ball—the 5.7-pound device had a 10.4-inch TFT screen, offering display power and real estate basically unheard of in a portable computer. The $4,350 machine also had a removable hard drive, a front-loading floppy disk (so you can avoid elbowing your seatmates on an airplane), and impressive horsepower for the time. *PC Computing* gave its annual "Most Valuable Product" to the 700C, calling it "a clear standout by its combination of speed, beauty, hard-nosed practicality, and, yes, grace."

Pierce continued,

> Everyone else seemed to agree. *PC Magazine*'s reviewer said it was "the best notebook I've ever used," and nearly every other review was equally smitten. But more importantly, the 700C was immediately in high demand. Bob O'Malley, then the managing director of IBM personal systems in Asia, told *Electric Business Magazine*'s Peter Golden that it had so much cachet that the tech people suddenly weren't making the buying decisions on notebooks. Now the CEOs were saying, "I want one of those, nothing else." The ThinkPad got IBM where they needed to go to expand all of their businesses—right inside the executive suite.

Pierce quoted my friend and boss Peter Hortensius, then an IBM researcher, as saying the 700C finally did what users wanted: "It really nailed something [the] ThinkPad has always tried to do, which is really solve user problems. A screen I can read on, a keyboard I can type on, a pointing device like the TrackPoint. . . . It was one of the first credible computers of that size, and it drove a lot of excitement early on."

"Sales, market share, and IBM stock soared," Pierce continued. "Thanks to the critical and commercial success of the 700C, Think-Pad became a powerful name almost immediately."

The excitement extended all the way to the American White House. President George H. W. Bush knew IBM Chief Executive Officer John Akers because they had both gone to Yale and both served as fighter pilots in the US Navy, although at different moments in history. Bush called Akers on December 18, 1992, to ask for a ThinkPad for his wife Barbara, who was going to use it to write a book about their dog, Millie.

It was the Friday before Christmas week, and Akers had already left the office to spend the holidays at his home in Nantucket. His executive assistant, Deanne Spencer, was desperate to figure out a way to respond to the president of the United States and ran down the hall to where Akers's young aide, Jim Steele, was stationed.

"Jim, you've got to take this call," she said. "The president is on the phone and he needs help."

Bush did not feel the need to reintroduce himself when he got on the phone with Steele; he simply started explaining: "Hey, I've been reading about this ThinkPad and want to get one for Barbara for Christmas. They are sold out. Can you get me one?"

"Yes, sir, Mr. President, we'll have a ThinkPad delivered to the White House," Steele promised with great confidence. "You'll have it in time for Christmas."

Bush said he was not looking for any deals and promised to send a check to pay for the machine. "I would have sent him mine for free," Steele recalls now.

Steele called the general manager of IBM's PC division, Jim Cannavino, and explained the situation. Cannavino took the next Think-Pad that came off the assembly line in Raleigh and shipped it to the White House. The IBMers knew that the White House had received the ThinkPad and that it was going through security clearance, but they never expected to hear directly from Bush again.

A couple of days after Christmas, Steele received a card from Bush with a note that must have been written just before Christmas. "Barbara will be thrilled with her Christmas present," it read. "Thanks for arranging it. Merry Christmas. George Bush." Included was a check made payable simply to IBM.

That was the power that the ThinkPad displayed within just a couple of months of its formal introduction. It became a status symbol for presidents and chief executive officers even if they were not computer literate. At that moment, it seemed to be powerful enough to possibly transform the image of IBM from an old stodgy company to one that could be vibrant and agile.

It was only after I saw these types of accolades and heard these kinds of stories that I understood what my team and I had created. The first ThinkPad won more than three hundred awards in a very short period of time. But there was no time for gloating. It was as if we had just fired the first shots in what promised to be a long war. The competitive battle was only just being engaged. There would be no vacations and no resting on our laurels.

Chapter 4:
The ThinkPad Goes to Space

When the space shuttle *Endeavour* took off from the Kennedy Space Center on December 2, 1993, to repair the Hubble Space Telescope, it carried a ThinkPad onboard. At the last minute, a young astrophysicist at the National Aeronautics and Space Administration (NASA) named John Grunsfeld had come up with the idea to load a ThinkPad with hundreds of images of various parts of the telescope, as they had been originally manufactured on Earth.

The astronauts would not actually take the ThinkPad outside when they made their space walks (EVAs), but they would be able to refer to the images inside the cabin in preparation for going outside. Or alternatively, astronauts inside the cabin could refer to the images and communicate with their colleagues outside in space suits. The mission was to install a new main camera and a corrective optics package on the Hubble telescope. That could only be done up close and in person.

More broadly, Grunsfeld wanted to see if a notebook could survive in space. The *Endeavour* had other IBM computers onboard to manage overall operations of the craft, but this was the first time that a notebook of this size and equipped with Intel's latest 486 chip had gone on such a mission. This chip was physically smaller than the previous generation and some experts felt that would make it more vulnerable to radiation.

It was well-known by then that one of the greatest hazards in space—to both people and the things they brought into space—is radiation. Altogether there are roughly thirty different types of radiation in space. Some of them are relatively harmless. When mild radiation passes through astronauts and their equipment, it usually has no lasting effect. Astronauts know when radiation has passed through the optic nerves in their eyes because it triggers flashes of light in their field of vision. Naturally, the further one goes into space, the more serious the radiation effects become. At the time, Hubble was located roughly 350 nautical miles into space, or about one hundred miles further into space than the International Space Station is positioned today.

Radiation that had large particles was dangerous to the ThinkPad. When the particles hit the semiconductors inside computers, they could form tunnels through the semiconductor material. That created what the experts called an "upset." One variation of that was a "bit-flip." If radiation hit the semiconductors that powered a computer, it could flip the computer's 1-0-1-0 binary code to the opposite, 0-1-0-1, which obviously was bad because all data would be corrupted. Radiation might also cause the computer to burn out completely. Journalists had fun writing about the prospects. One reporter for *Government Computer News* wrote that NASA wanted computers that did not "fry when they fly." And a reporter for the *Miami Herald* wrote that the ThinkPad was a "guinea pig."

The liftoff of the 250,000-pound *Endeavour* from Florida at 4:27 a.m. that very clear December night proceeded as scheduled, and the ThinkPad operated normally on the first few days of the mission, which was numbered STS-61, the sixty-first scheduled space shuttle mission. The Hubble repairs and refurbishment went well, and the astronauts did not need to refer to the ThinkPad.

But midway through the eleven-day mission, astronaut Kenneth Bowersox, who was piloting the shuttle, attempted to power up the machine. It did not respond. He suspected that radiation had passed through it and crippled it. The suspicion was immediate and obvious: ThinkPads could not survive in space.

Winning the Deal

Because of my interest in space as a teenager in Nagoya, I was very pleased one day in September 1992 to get a call from Andy Klausman in the United States. He was IBM's key liaison with NASA, and this was the first time I had ever spoken with him. The first Think-Pads were being introduced, and he had heard about what we in Yamato were trying to achieve. The formal launch was scheduled for October.

Andy was excited and spoke rapidly. He tried to explain that sending any notebook into space would require it to operate in one G on earth, but then three Gs or four Gs on blastoff and finally zero Gs in space. Could we do that? I had no idea what he was talking about because he was talking so fast and I wasn't familiar with the use of the letter "G" for "gravities." What he was asking was whether the machine could handle takeoff and the weightlessness of space.

When I figured out what he was asking, I told him that no, we had never tested the machine for those special requirements. Why would we? We never imagined that the ThinkPad would be operated anywhere but on Earth.

My immediate concern was whether or not the hard drive would work in space. This is why: If you are old enough to remember playing records on a turntable (this goes back to the era before MP3 players or iPhones), you recall that a stereo system's needle created music by "reading" the electrical audio signals in the grooves of the record and amplifying them. A hard drive was very similar to the record player. Just as a needle floated lightly over a record, the hard drive's read-write head also depended on a certain amount of air and gravity to maintain the right position in relationship to the disk.

We started doing some testing and arranged to get an early prototype of the 700C to the Boca Raton lab. Andy flew from Houston, Texas, to retrieve it and started pitching its qualities to his NASA contacts at the Johnson Space Center in Houston. "When they saw it, they said, 'Wow, we have to talk,'" he recalls. Actually seeing the smaller dimensions of our ThinkPad was powerful.

Andy, a software engineer still working today in the space industry in Houston, had been involved in developing the flight software for the shuttle, so he was deeply knowledgeable about the agency's thinking. NASA used five of the company's AP-101 line of mission-critical aviation computers as their general-purpose computers to fly the shuttle and control its onboard systems. This line of computers was used in B-52 and B-1B bombers as well.

We also made what NASA called data electronics units that displayed flight information to the astronauts. But for years, for other noncritical computing purposes, NASA relied on GRiD Compass computers that were called portables but were twice the size of our ThinkPads. The original onboard computers, made by GRiD Systems Corp., had their own operating system and required customized software. They couldn't operate on batteries. Their orange electroluminescent flat-panel display screens were hard to read. In short, the company was not keeping pace with the new technologies that were rapidly becoming available. "A lot of the stuff that NASA wanted to do, they couldn't do it with that machine," Andy said.

Eventually, GRiD started to fail as a business and was acquired by AST Research, Inc. All of which created an opening for us.

Our Inside Man

Andy's key contact inside NASA was John Grunsfeld, who had just arrived in Houston from the California Institute of Technology (CalTech) as an astronaut candidate. As I later discovered, Grunsfeld had an incredible background. He graduated from the Massachusetts Institute of Technology with a bachelor's degree in physics. He earned a doctorate in physics from the University of Chicago using a cosmic ray experiment on the space shuttle *Challenger* for his doctoral thesis. He then joined CalTech as a senior researcher in physics, mathematics, and astronomy. Grunsfeld understood the intersections among these different disciplines, which equipped him well for the era of space exploration that was unfolding. For now, he was assigned to Mission Support in the Astronaut Office. This office was the key liaison between astronauts and the rest of the sprawling NASA bureaucracy. Grunsfeld considered himself a "hunter-gatherer" whose job was to scour the civilian arena for the latest technologies and quickly get them into space.

At CalTech, Grunsfeld and other scientists used something called NSFNET, which stood for National Science Foundation Network. The NSF, an arm of the US government, created the network to link key researchers at academic institutions. Scientists could scan the papers written at institutions around the world for the latest research. Most accounts of the origin of the Internet point to the key role of the Advanced Research Projects Agency Network (ARPANET), which was created by the US Department of Defense, but it was not alone. NSFNET, ARPANET, and others were ultimately merged to create the Internet, which was still very slow in the early 1990s. America Online (AOL) introduced its iconic "You've Got Mail" in 1989, but it was still awkward to dial in via a telephone modem, and computers were still too slow to give the Internet truly mass appeal.

But Grunsfeld had witnessed the power of computing networks at CalTech and was a believer. NASA as an institution, however, had not yet bought in to that vision of the future. It was ironic that an institution that depended so heavily on technology was skeptical about computers, at least initially. Grunsfeld recalls that he was "amazed that computers were viewed as unnecessary accoutrements." NASA's top management, made even more cautious by the *Challenger* disaster in 1986, did not think astronauts needed computers. The prevailing view was "If we ran the Apollo program without astronauts having computers, why do they need them now?"

Consequently, the Astronaut Office had only primitive email connectivity. "I think there was one Intel 286 personal computer for every six or eight people," Grunsfeld explains. "We shared it to do email." Because of his background, Grunsfeld emerged as the leading IT specialist for mission support. From our perspective, he was the right man in the right position.

Grunsfeld was determined to get a modern notebook into space and looked at many different models. Ultimately, he was drawn to the ThinkPad because of its TrackPoint. Grunsfeld knew that a mouse would not work in space because it would float away. Trackpad concepts, which required the user to apply pressure to a pad, also would not work well in space because by an astronaut applying pressure in this way, his or her body would be forced to rise up and away from the computer. With an astronaut's hands poised on the edge of the ThinkPad, the TrackPoint could be manipulated better with less physical force than any other interface device.

As Andy engaged with Grunsfeld and others during the rest of 1992 and into 1993, it became clear that NASA was interested in the ThinkPad because its body was made mostly of carbon composites, unlike other notebooks whose bodies were constructed of metal. They thought that carbon would not melt or combust if a short circuit occurred while in the confined space shuttle crew cabin. Any such incident obviously would threaten the lives of astronauts. In fact, little did we know that NASA actually burned one of our ThinkPads to test for harmful gases. It found none.

One other worry was that microscopic particles of metals were inside the machines. In normal Earth conditions, those metal particles never moved. No one was ever aware of them. But NASA scientists and engineers were worried that in a low-gravity environment the particles could float and cause electrical shorts inside the machines. So they insisted on applying a nonconductive coating material to the innards of the machine.

Another issue was that heat did not dissipate in space the same way it did on Earth because there was no gravity to pull it away. So we had to make sure the fans would prevent the ThinkPads from overheating. But other than these relatively minor modifications,

NASA was getting machines that were the same as ThinkPads we were making for use on Earth. That's what NASA wanted—they wanted to buy commercially available "off the shelf" computers rather than creating their own, which would have been infinitely more expensive. NASA also wanted notebooks that the astronauts could practice using at their desks on the ground while training for a flight.

At first, the ThinkPad's opportunity to get inside the shuttle was a proposal to use notebooks to practice landings while in orbit. The problem in previous flights was that during the two weeks in orbit, the crew commander would not have the opportunity to practice landing and might get rusty.

The advantage of our machine was that our screen resolution was so high, compared with that of other portable devices, we thought we could simulate the view out the windows that the crews would have during a landing—and do so in color.

We kept talking to NASA, and they kept testing our machines. While Andy was showing the 700C to NASA, we already were working on new iterations of the ThinkPad. The next model was going to be the 720C and then the 750C. This last one, the 750C, was going to use Intel's 33-megahertz 486SL microprocessor; it would be much more compatible with other systems and would have sufficient processing clout. We were pushing ourselves to win this business. Much as Harvard Business School had galvanized us on to create the L40 SX and the first ThinkPad, the prospect of getting ThinkPads onto the space shuttle drove us very hard.

We in Yamato were coordinating very carefully with Andy, and it finally worked. NASA made it official and ordered the 750C—but only for the very limited purpose of training, including practice landings. My concerns about the hard disk drive failing in space had proven unfounded. The ThinkPad would not actually be used during a blast-off when it was being subjected to two-G or three-G conditions. It would be operated only in a pressurized cabin. The hard disk drive functioned perfectly in those conditions.

The Moment of Truth

Grunsfeld knew it would take a long time to get the notebook through all of NASA's certification processes. There were many steps in that effort involving different panels and review boards. It was particularly difficult when NASA was evaluating an off-the-shelf commercial product because it would have to take the piece of equipment apart and analyze every component. Often, the agency would test a product for so long that it was overtaken by newer commercially available technologies. It had taken NASA twenty years, for example, to move from mechanical instruments used to control its vessels to display screens.

Grunsfeld knew that the ThinkPad would be safe if it burned because of the previous test that had been done. He also knew that turning it on would not interfere with the shuttle's computer and communications systems because of a separate test that also had been performed on that issue. One test the ThinkPad had not been subjected to, however, was the radiation test. To do that, it would have to be sent to a massive particle accelerator and bombarded with radiation.

One day, as the launch date approached, Grunsfeld saw an opening. It involved the pictures of the telescope that the crew was supposed to take along with them—thousands of eight-inch by ten-inch photographs. The crew was supposed to sort through them and decide which ones to carry, which would then be stored in ziplock bags, and the crew would have to create an index scheme to find the right image at the right time. For the solar array replacement, they had to find photographs of the solar arrays and the cables and connectors. They had a similar folder for gyroscopes. In the end, they selected enough photographs to fill two lockers on the shuttle, which was a lot of valuable real estate. "I thought what we really needed was a killer app for that crew," Grunsfeld recalls.

That killer app would involve scanning the photographs into two of the largest hard drives they could find. They could be inserted into the ThinkPad as needed. This was no easy task because there were none of today's JPEG compression files. In those days, people used dial-up CompuServe modems and that company's Graphics Interchange Format (GIF) to share photographs. GIF could support only eight bits per pixel, a resolution that is primitive by today's standards.

One day, pilot Ken Bowersox went to the Astronaut Office to meet Grunsfeld. Sox, as he is widely known, was an astronaut from 1987 to 2007. He was going to be the pilot on the mission to fix the Hubble telescope. Although he would not be directly involved in the repair work, he shared Grunsfeld's sense that it was time to embrace newer technologies.

"Time was running out" for figuring out how to manage the photos, Bowersox recalls. "The crew was way too busy to handle anything like that themselves. So I went to the Astronaut Office and talked with Grunsfeld."

"What if we fly one of these ThinkPads?" Grunsfeld asked.

"That's fine with me," replied Bowersox, "if we can get it approved."

The NASA bureaucracy had not yet fully certified the 750C for crews to use, but Grunsfeld said the key tests had been positive. Bowersox agreed and proposed to the crew that they bring the ThinkPad. They also agreed. It was an end run around the formal approval process, which could have taken years longer. The goal was noble—to do everything possible to help the crew on the *Endeavour* achieve success.

Here's how Bowersox describes what happened when he took the 750C into space:

A lot of people thought the radiation would be too strong for the ThinkPad. It had hit the GRiDs earlier.

The TrackPoint worked very well in orbit. A lot of the earlier computers didn't have that. That made the ThinkPads user-friendly from our perspective because we were working with our bare hands inside the module. You didn't have to position yourself as finely, and you had finer control of the cursor with the TrackPoint. We did not use Microsoft Windows because we were concerned about its stability. When you're in space, you want to make sure everything works all the time. We stayed with Microsoft's Disc Operating System (DOS) 6.

The problem was when you got into radiation zones, the personal computers were one of the first things to die. There were certain zones that were bad like the South Atlantic anomaly near the coast of Argentina, which is where the Earth's radiation belt comes closest to the planet's surface. High energy particles in that area caused a lot of upsets.

Every once in a while you'd get a hard crash of the battery pack. It looked like radiation had done it. We called these single-event upsets because all it took was for a charged particle to hit a microprocessor and it would interrupt the operation of that device. It was just random.

On STS-61, we would check the ThinkPad every day and then one day it was just dead. I had to take it completely down. Just power cycling the switches wasn't enough. It looked like it was a problem with the battery pack. I disassembled it and took the battery pack out of it and the hard disk drive.

Then something happened. "When I put it back together, it worked just fine," Bowersox recalls. "I was surprised." It was as if the ThinkPad had come back from the grave.

The Outcome

Altogether, the 750C's memory (its semiconductors) had been hit eighty-five times by radiation—and survived. Its memory had recorded the hits and the precise times they occurred.

Andy put out a press release on February 21, 1994, that read in part:

During the December 1993 Space Shuttle mission to repair the Hubble Space Telescope, an off-the-shelf IBM ThinkPad 750C color notebook computer successfully flew into space for the first time. . . . Testing on the ThinkPad provided more than 202 hours of data. The returned data indicated that memory had been changed 85 times during testing. The radiation never prevented the ThinkPad from properly functioning and performing its tasks.

No one except Bowersox knew about the near-miss until STS-61 returned to earth. Engineers later determined the cause of the problem was that radiation hit the ThinkPad's power system, but not severely enough to permanently cripple it. The bottom line was that the 750C had survived.

NASA was so pleased that new doors to the ThinkPad started to open. It agreed to use ThinkPads on future flights to monitor payloads and transfer data between the space station and Mission Control at high speeds, which enabled two-way video communications for the first time. It wasn't very long before astronauts, while still in orbit, could watch video recordings of their vessels as they blasted off or have video conferences with their families. The video from the ground could be fed to them in real time on their ThinkPads.

Another use in the shuttle was the display of orbital information. The Pacific Ocean is vast, even when seen from space, so the Think-Pads became useful in displaying maps that identified precisely where the astronauts were as they peered down on Planet Earth. Those maps changed as the shuttle moved, much as the video-screen maps on seatbacks in today's airliners do.

NASA eventually replaced all the GRiD computers on the shuttle with ThinkPads. Since 1994, there has not been a single human space flight without a ThinkPad on it. Our early victories with the space shuttle set us up to provide ThinkPads to the International Space Station as well, a few years later. It was a great convergence of technologies, a golden age. All of us felt lucky to be part of it.

Reality on Earth

Louis V. Gerstner Jr. had become chief executive officer of IBM in April 1993 and launched one of the most painful but perhaps necessary transformations of the company we all loved. He decided to reduce the company's employment by thirty-five thousand after his predecessor Akers had laid off forty-five thousand people the year before. The company was on track to lose $16 billion in 1993. Further layoffs appeared inevitable. It seemed the company just could not adapt fast enough to changing technologies. It was too big and its culture was too insular.

To boost the morale of the employees who remained, IBM asked Ken Bowersox to fly to its headquarters in Somers, New York, in the spring of 1994. Andy joined him and brought the computer that was flown on the shuttle in a special travel case. Aside from appearing at an employee event, the two of them met briefly with Gerstner.

"It was just for five minutes," Bowersox recalls. "I asked him if he had any advice for NASA. He said, 'What I tell everybody is remember, it's your people. You have to take care of them. You have to have a concrete objective for your people.'"

That might have been one of the most ironic moments in IBM ThinkPad history. Andy's business unit, the IBM Federal Systems Company, had already been sold to Loral as of January 1, 1994, for example, so he felt odd going to IBM headquarters as a former employee. One of the most radical restructurings in the history of IBM was underway. But we at Yamato were still standing. In fact, we were growing.

The ThinkPad Becomes NASA's Workhorse

From Japan, it was impossible to see NASA the same way Andy Klausman and other IBM people involved in the NASA account could.

They shared information with us on an "as-needed" basis. We were asked to create different tests for our machines and to make mild modifications over the years, but we never fully understood what was happening inside NASA.

Now, in retrospect, it's clear that Grunsfeld became the leader of what can only be described as a ThinkPad movement within NASA. A veteran of five space missions, he took the lead in mapping out a strategy for how ThinkPads would be used in orbit and on the International Space Station. He would rise to one of the most senior positions in NASA as assistant administrator in charge of science and astronauts.

After the ThinkPad demonstrated that it could survive in space in 1993, Grunsfeld's first journey into space was as a mission specialist on STS-67 in May 1995. This mission was heavily focused on using three space-based far ultraviolet astro-telescopes (Hubble relied on different types of light) to explore the far reaches of space. The space-based telescopes could see more clearly than ground-based telescopes because the Earth's atmosphere interfered with their vision. Some GRiD computers were still on board STS-67, but Grunsfeld was able to persuade NASA to let him bring ThinkPads as well, and he wanted to demonstrate two new uses for them.

The first was to figure out a way that the astronauts could see the results of what the space telescopes were discovering, rather than the results being communicated only back to NASA on Earth. Grunsfeld and fellow specialists were able to figure out a way to run the scientific analysis software on board on a ThinkPad as data came in from the astro-telescopes. They could see what the telescopes were seeing.

The second objective was more ambitious. The astro-telescopes were pointed using star trackers and a special automated software program. This system used powerful cameras to search space for specific patterns of three bright stars, which served as a reference point for where the telescopes should be aimed. In the parlance, the star trackers and software were supposed to "acquire" a target. Various errors could throw off the success of the target acquisition and, without that, there could be no discovery. Some of the galaxies to be studied were extremely faint in the ultraviolet light and were billions of light years away. One light year is 6 trillion miles. The distances were beyond the imagination of most humans. Finding a galaxy could be like trying to find the proverbial needle in a haystack.

The star tracking software had not performed up to expectations on a previous mission and much science was lost. So Grunsfeld concocted a solution. Before blasting off, he had worked with engineers at the Johnson Space Center to build a small box of electronic equipment that intercepted the data stream coming from the star trackers. Previously, that information was available only to the shuttle's computers and was sent to the ground without astronauts being able to see it. Now astronauts could see it. "We intercepted the star tracker data because I was convinced that if the automatic acquisition wasn't working, the human brain is a really good pattern recognizer," Grunsfeld explains. "We flew a computer program on the ThinkPad that showed a map of the three brightest stars around the object we were looking for and then we intercepted data that gave the position in X-Y coordinates of the three brightest stars. We had a little joystick, a hand controller. We could move and rotate the telescope until the stars aligned in a triangle with what we could see on the map. It was pivotal to the success of the mission that we had this capability. We ended up using the ThinkPads a lot."

In short, Grunsfeld had used the ThinkPad to help probe the deepest corners of space.

Back on Earth after that mission, Grunsfeld continued his ThinkPad campaign. "I convinced management that in order for astronauts to be proficient in using ThinkPads in space they should also have ThinkPads in their daily work," he recalls. "The computer that every astronaut worked on every day should be a ThinkPad."

The number of ThinkPads in use inside the space program shot up in the 1997–1998 time frame to several hundred. The agency created a culture in which astronauts carried ThinkPads with them basically everywhere they went. It wanted them to know how to change network settings and operate the machines with a high degree of proficiency because there was no IT help desk in space. If astronauts had a problem with a ThinkPad in space, the chances were that they would have to resolve it by themselves. "I wanted the ThinkPad to be the currency that everyone had," Grunsfeld says.

The second time Grunsfeld went into space was on STS-81 in 1997 on the space shuttle *Atlantis*. This was a ten-day mission to dock with Russia's Mir space station. By now, we at Yamato had come out with a new ThinkPad called the 755C, which had a faster Intel chip. It was powerful enough to operate software that helped pilots plot the most fuel-efficient routes to make a rendezvous with Mir or another space station. It was called the Rendezvous and Proximity Operations Program (RPOP). Pilots even competed with one another to see who could make the most efficient rendezvous dockings. Astronauts also could now use their ThinkPads to increase their situational awareness of the vessel's robotic arm and its movements, which helped improve safety.

The Russian cosmonauts on Mir also were ThinkPad fans. But they had only the old 750Cs, not 755Cs. (One of the Russian 750Cs is likely to set a record for time spent in space, according to *SpaceRef*, a space news and information service. It was left in Mir's damaged, and later abandoned, Spektr module. It will float in space forever.)

After the Mir docking mission, Grunsfeld was asked to start creating a plan for an information technology infrastructure for the International Space Station, the first pieces of which were sent into space in 2000. Grunsfeld had big ambitions to use ThinkPads and other computers to build a network that would be both wired and wireless. "Management said 'Never,'" he recalls. But he persisted, as usual. "The ThinkPad really became part of the DNA of everything we were doing for the space station."

Grunsfeld decided to link the ThinkPads on the station wirelessly with each other and with Earth, which he said "was good because you could grab one and float around." The astronauts used their Think-Pads to organize video conference calls with one another and with NASA specialists back on the ground to solve work-related issues.

We, in Japan, knew about one other use of the ThinkPads on the space station. The machine was equipped with a DVD drive. Bowing to the wishes of the movie industry, we manufacturers had placed "region locks" on laptops and other devices that prevented a user, say, in China, from viewing a movie recorded in Europe. But now, NASA asked us to create "region-free" DVD drives. Our conclusion was that astronauts from different nations and different continents were spending long periods of time on the space station and wanted to watch movies from their home countries. The ThinkPad was not only helping with their work but was also helping to entertain them.

The space station was finished in stages by 2011. "I think the ThinkPad was a more important tool than almost any other tool we had on board," Grunsfeld concludes. It is still in use today.

Chapter 5:
A Butterfly's Day in the Sun

We in Yamato had won the right to control the engineering of the ThinkPad 700C and its next iterations such as the 750C and the 755C. The machine was clearly oriented toward the business and educational markets plus specialized sectors like NASA's efforts in space. We did not pretend it was right for the mass-consumer market. Because of its many advanced features, it was too expensive for the average consumer—and we certainly had no interest in reducing the price.

By now, the design lab in Boca Raton had been closed, and all of that facility's activities would eventually be redistributed to Raleigh and elsewhere. That left Raleigh as the only other place in the personal computer division where a product could be designed and engineered.

It was Tim Cook, later the chief executive officer of Apple, who first argued that IBM should try to devise a notebook tailored to everyday consumers. I never had any direct contact with him and did not even know of his role before conducting research for this book. Cook, an industrial engineer with a master's in Business Administration (MBA), worked his way up the IBM ladder to become North American fulfillment director for the PC division. Based in Raleigh, he managed manufacturing and distribution throughout the Americas. Altogether, he spent twelve years at IBM. Not surprisingly, Apple did not respond to a request to make Cook available to comment.

Cook and fellow engineer Mike Malpass, who was my source for much of this account, had worked together in the mid-1980s when Cook was in charge of manufacturing monitors for mainframe computers. Cook was trying to drive down the cost of making the monitors, which were very labor intensive to produce and therefore expensive. He worked with Malpass, who was a lead mechanical and electrical engineer, on the installation of an automated assembly line.

Cook's and Malpass's common foes were the monitor's development engineers, who did not care very much about how difficult it was to actually make their product. Manufacturing it was someone else's problem. The "silos," or separate fiefdoms, inside IBM were very rigid. Malpass says "it was like pulling teeth" to persuade the development engineers to change their designs so they could be assembled more easily, but Cook and Malpass prevailed. By the time it was over, robots could assemble the monitors with just five screws. It was a successful collaboration between the two men.

Their careers took them in different directions until 1990, when Cook asked Malpass to join him, this time in the PC division. The problem once again was complexity. Cook asked Malpass to assemble a group that would simplify the design of existing IBM personal computer products.

They launched a project code-named Stealth because it would be "under the radar." The top management of the North Carolina operations of IBM knew about it, but they did not tell top management of the PC division in New York or anyone else at senior levels of the company. It was not terribly unusual for someone down deep in the trenches at IBM to try to develop a concept or a product without informing top management. The presumption was that top management, preoccupied with budgets, would kill anything that was unproven, yet another indication of the dysfunctional culture that gripped the company.

Cook and Malpass's team of about thirty-five people developed and learned to manufacture a lower-end notebook. They unveiled it to the head of the PC division, Robert Corrigan, when he made a visit to Raleigh to discuss other matters. He walked into a room and there sat the notebook on a table. Selling him an actual product was easier than it might have been to sell a still-untested concept. Corrigan accepted it, and it came to be known as the PS/Note. It was made in the United States, which was one of Cook's goals, and it retailed at less than $2,000. It was branded as an IBM product, but not a ThinkPad.

That success encouraged Cook and his team to try to build on those relatively humble beginnings and design a higher-end notebook for consumers. The company agreed and funded the project, which was to be code-named Butterfly. We in Yamato knew about it but weren't directly involved.

So as Yamato was rolling out the ThinkPad 700C and its successors, Cook's group was now officially developing a notebook aimed at the consumer market. The key issue facing them at the time was that the biggest LCD screens were 10.4 inches measured diagonally. But the Cook team wanted a full-size keyboard measuring twelve or thirteen inches across. That meant the keyboard would be wider than the screen, which was impossible to achieve in the traditional clamshell form factor.

Cook had heard about a keyboard that had been designed at IBM's research division in Westchester County, New York. It was conceived of by one of my friends, the late John Karidis. He was one of the very best mechanical engineers I had ever met.

Karidis had come up with the concept of cutting the keyboard into two rough triangular pieces along an uneven path so that when the computer was opened, the left half of the keyboard would move to the left and the right half of the keyboard would move to the right. That created a complete keyboard that actually hung over the edges of the base of the computer, creating the appearance of a butterfly, hence the name. The physical movements of opening and closing the computer caused the keyboard to expand and contract. I thought it was a brilliant idea and was excited about it, even though it was going to take a great deal of engineering work to turn it into a reality.

It was my policy that whenever a new group within IBM tried to come up with a new product, I would help them even if the new group could emerge as a potential competitor of the Yamato lab. After all, I had helped the Boca Raton lab when they had such troubles with the PS/2 back 1987. There were times to compete and then there were times to cooperate. Top management wanted us to cooperate with Raleigh in developing the Butterfly and for them to cooperate with us in improving our ThinkPads for the business market.

I tried to encourage my engineers to exchange technological details with the Raleigh team. We had many long meetings, days and days of meetings, but there was never the open and robust exchange of ideas that top management wanted. Both sides were too guarded. In the end, we learned some things from them and they learned some things from us. Cook never took part in these meetings, however, and I never at any point met him personally.

Meanwhile, Malpass and others were starting to build excitement within IBM in the United States for the Butterfly, which was called the 701C. Top management invited Malpass to bring a wooden prototype to an off-the-record session with top technology writers in New York. The journalists had signed nondisclosure agreements. They wanted a peek into the future and IBM wanted to see how they reacted to different ideas being developed. They went crazy for the Butterfly. Each of them wanted to hold it and see how it worked.

The Butterfly team was two to six months late meeting the product launch schedule. No one agrees on just how late they were, but they were clearly months late. The original launch date was in 1994. Cook had left IBM earlier in 1994, so he was no longer part of the equation (he joined Apple in 1998) and Malpass also left the Butterfly group for family reasons before it launched. Those management shifts may have contributed to the Butterfly's delays. Another problem was refining the mechanism that physically moved the "wings" into place. The mechanism had to survive tests in which it would be opened and closed thirty thousand times.

The product did not go on the market until 1995. Despite that, it was a hit. It was the top-selling notebook of the year. It had the most first-day orders of any notebook ever made by IBM up until that moment.

The Butterfly truly captured people's imaginations, as it had mine. One IBM television commercial showed footage of the F-14 *Top Gun* jet fighter, whose wings moved when the aircraft landed, suggesting, however whimsically, that the Butterfly was like an F-14.

James Bond (played by Pierce Brosnan) used a Butterfly very briefly in the 1995 movie *GoldenEye*. The Museum of Modern Art in New York loved the Butterfly so much that they added one to their permanent collection of well-designed objects. Overall, the machine received twenty-seven design awards.

But the delays had hurt the Butterfly and top management's confidence in the Raleigh team. Because of the unexpected problems Raleigh was having, the company decided to concentrate all ThinkPad development in Yamato. Bruce Claflin, who had been general manager of the PC division since the first ThinkPad was introduced, came in person to visit me in Yamato. It must have been in late 1995. We already had introduced two new iterations of the original 700C, namely the 750C in 1993 and the 755C in 1994. We were getting things done.

"Naitoh-san, I know the Butterfly keyboard is not made here," Claflin said. "You may not like this, but take this one as your adoptive child."

In other words, I had to decide how to manage the product. I didn't hate the Butterfly concept. I loved it. It had been created by my friend. Claflin did not have to tell me to love it. I already did.

But because of the delays, the Butterfly used the Intel 486 chip even though Intel was introducing its much improved and much faster Pentium processors. The machine was thus already a generation old in terms of processing speeds. And the size of LCD screens kept getting bigger. By the time the Butterfly hit the market, the latest screens were more than twelve inches measured diagonally. That eliminated the need for a Butterfly keyboard. "We cracked the code on making a larger piece of glass (for the screens), so the need for having this expanding keyboard became much less," design chief David Hill told *ComputerWorld*.

I went to North Carolina to meet with members of the team so that I could create a strategy to keep the Butterfly going. A series of meetings were scheduled for late in the afternoon over a period of days. When I went to the assigned room, nobody else showed up. The other invitees already had gone home. I don't know whether they had a sense of failure or whether they blamed Yamato and me for not transferring enough technologies to help them. I felt sorry for them. Any team could fail.

With regret, I made the decision to suspend the program. In my view, any rivalry between designers in Yamato and Raleigh was not the decisive factor. It was a business and technology judgment. "The Butterfly had a brief window when it was aligned with people's desires," Hill said.

Here's how *Business Week* described it in February 1996: "IBM has targeted its nifty Butterfly laptop for extinction. On the market barely a year, the ThinkPad 701C is a hit, but executives feel it will soon be outmoded. Poor Butterfly," the magazine continued. "It's doomed because the next generation of IBM lightweight PCs will sport screens wide enough to allow a regular keyboard."

We in Yamato loved the Butterfly idea so much that we created a toy model–sized version of it and gave it away as part of our marketing efforts on the tenth anniversary of the ThinkPad's introduction. People could assemble their own Butterflies just like they could make a model car or a model airplane. We later tried to use the Butterfly concept in other products, but it never worked. The Butterfly's moment in the sun was gone forever. And we in Yamato were the last ThinkPad design and engineering team left standing.

Chapter 6:
The Quality Battle:
How American Students
Punished Our Machines

When I think back to the beginnings of the quality challenges we faced, one image remains etched in my mind. We had designed the early ThinkPads on the assumption that they might be opened and closed ten times a day. Of course, we had no idea of how different customers would use the machine, but that was our estimate. Some early customers, however, opened and closed their machines many more times, putting pressure on our hinges. That, in part, resulted in our high failure rate from broken hinges in those days—about 19 percent of the machines produced were returned with problems. We clearly had to do something to reduce that rate.

In 1996, Ken Yonemochi, who was head of mechanical engineering and therefore the point person for quality, went to Australia to spend time with the Coca-Cola sales people who had purchased ThinkPads. Coca-Cola was one of our earliest and biggest corporate customers.

When Yonemochi went on sales calls with the Coca-Cola people, he quickly recognized that they were opening their ThinkPads and booting them up dozens of times a day as they paid calls on different customers. This was far in excess of our estimate and put tremendous stress on the hinges.

When he returned and explained the problem, it became clear we needed to figure out a way to better test the hinges and begin assembling data so that we could make them more durable. We decided we needed to open and close a machine ten thousand times to start the process. But we didn't have any machinery that could do that.

So Yonemochi tapped a young mechanical engineer on his team, Hiroyuki Noguchi, and assigned him the task of opening and closing a ThinkPad ten thousand times in rapid sequence. The young man undertook the assignment with stoicism and bravery.

I remember leaving the office that night at my usual time of about eleven. All the lights were off, except for the light at Noguchi's desk. He was sitting there opening and closing a ThinkPad with great devotion. It was going to be a long night for the young man. He would see the sun rise.

Clean Sheet

It may seem like a paradox, but it is nonetheless true—the ThinkPad quickly developed a reputation for technological excellence and reliability, yet at the same time, we were coping with enormous quality problems. The business was growing very fast and we had multiple products on the market by the 1995–1996 time frame, but customers were writing letters to IBM CEO Gerstner complaining about failures. The chief information officer of one major customer, Chevron, called the machines "StinkPads." A paraplegic user wrote to Gerstner and complained that he had suffered third-degree burns on his lap from the heat generated by a ThinkPad. He couldn't feel his legs so he did not realize he was being burned. An angry Gerstner pushed us to improve. The entire industry was coping with very similar problems, but that was little consolation.

Adalio Sanchez, who became vice president of development in March 1996 and who would become general manager two years later, was instrumental in improving the way we all did things. He imposed deadlines and demanded constant reviews and constant progress reports. We developed closer connections with vendors and started meeting more frequently with customers to hear about the problems they were having.

As general manager beginning in 1998, he decided to attack one of the challenges we had been encountering, which was maintaining consistency in the small details that defined how a customer experienced our product. Where was the power switch located? How big were the various ports and where were they located? If a customer had an optional device that he or she used with one ThinkPad, what would happen when we introduced the next model? Perhaps that device would no longer be compatible. We wanted to maintain as much consistency as possible, but the technology kept changing and our machines changed probably a little bit too much.

Sanchez told us it was time to simplify our product line. We could start with a "clean sheet" of paper, in his words, to imagine how to improve the customer's experience. We would not have to maintain backward compatibility with previous iterations of the ThinkPad. We could simply concentrate on what was best for the customer. This would result in the second major generation of ThinkPads.

We were working on the clean sheet approach when Gerstner once again turned up the heat. One evening in March 1999, Gerstner hosted a small corporate dinner in Florida and Sanchez was invited. About six IBM executives were present. Sanchez had been sending Gerstner a new ThinkPad whenever a new model was introduced. Gerstner thanked Sanchez for the steady stream of ThinkPads but then immediately started complaining to Sanchez. Why did he have to relearn how to turn the power on or how to lift the lid every time he got a new machine? Where was the consistency? "Gerstner tortured me at that dinner," Sanchez recalls.

Having our own CEO complain about what we were doing came as a major jolt. We accelerated our work on the next line. The marketing people in Raleigh came up with the idea of naming the models with specific letters to make it clear which model was designed for which group of customers at which price point. Now we would have the X, meaning "extra light and extra small." The X was an ultraportable full-featured notebook slimmer than a deck of cards and lighter than a half-gallon of milk. At about the same time, we came out with the T, which stood for "thin and light" notebook for travelers. A little bit later, we came out with the R and the A. The R stood for "reliable." We thought all our machines were reliable, but this one had extra reliability built in. The A was an "alternative to a desktop computer."

Collectively, this was the XTRA generation of ThinkPads. In the naming of these computers, we moved from a BMW style of using numbers (the 3, 5, and 7 series) to a Mercedes-Benz alphabet style (C, E, and S.)

Let the Real Punishment Begin

But our quality challenges intensified in some ways as we expanded our sales beyond the corporate sector and into the educational world. After we started rolling out these products in 2000 and 2001, we soon faced a good news–bad news situation. Sales of the T series, in particular, started exploding in American universities, private high schools, and even some parochial middle schools. The T series came to be seen as the highest-end ThinkPad because of the way it combined all the best features and because it was very light. Even though we had been selling ThinkPads at Harvard Business School, most other universities were slower to adopt them.

But as a result of large numbers of sales in schools, our failure rates increased to more than 20 percent in some cases. The quality problems we had been fighting since the first ThinkPad was introduced and largely winning were back. Hard drives were crashing.

Keyboards were breaking. LCD screens were getting damaged. The motherboards that contained all the chips and circuitry were sustaining microscopic cracks and therefore no longer functioned. The whole industry experienced similar problems as students started using notebooks. We were not alone.

Our colleagues in the ThinkPad division in Raleigh told us that American students were using our ThinkPads in ways that we could not imagine. Sam Dusi, who was in charge of product marketing, told about getting called to a university that was having too many problems. The TrackPoint in the middle of the keyboard was making a dent in the screen, he was told. He knew that was impossible because there is a two-millimeter gap between the keyboard and the screen when the machine is closed.

Then he went to the Student Union and saw students gathered on the steps. It was winter and they didn't want to sit directly on the steps. So they sat on their backpacks—with their ThinkPads inside. Their weight was compressing that two-millimeter gap and causing the TrackPoint to jam into the screen!

We trusted our colleagues in Raleigh, but we just could not believe what they were saying. It did not make any sense. To us, the ThinkPads were precious. Why would American students intentionally abuse them?

We dispatched different engineers to study the problem, but our most decisive action was sending a team of engineers from Yamato including Kishiko Itoh and Takane Fujino to go see for themselves in 2004. They went on a tour of Wake Forest University and the University of North Carolina, both in the state of North Carolina, and of Bergen Catholic High School, Seton Hall University, and Union Catholic High School, all in the state of New Jersey.

What they saw astonished them. At both UNC and Wake, they saw walls of dozens of ThinkPad keyboards that had broken. Students who got bored in the classroom took pencils or pens and popped off the tops of the keys. Then they found they could not replace them. Or they would pound on the keyboards with incredible force as they played action video games, damaging the up, down, left, and right arrow keys.

However they had been broken, the students returned the machines and their keyboards to the help center the universities had established. We didn't want the broken keyboards back because there was nothing we could do to fix them in any economically viable way. So student workers at the help centers nailed the broken keyboards to the wall as a kind of stunt. They thought it was artistic. Our engineers were horrified.

Our engineers also witnessed students stuffing ThinkPads into their backpacks with heavy textbooks and then bicycling across campus. The backpacks weighed thirty or forty pounds. Fujino had been trying to create tests back in Yamato that would inflict the same pattern of damage he was seeing on broken machines; he couldn't understand how so many parts were being subjected to microscopic cracks. He studied them through a microscope and tried to imagine what could cause the damage.

Now the evidence was right in front of his eyes. By putting the ThinkPads in backpacks with other sometimes sharp objects like AC adaptors, bracelets, or keys, and then pounding the ThinkPads with heavy books in a confined space—at the same time that the bicycle ride was generating bumps—the students had created the ultimate nightmare for any notebook computer.

Another pattern shocked us: If a student's machine were plugged into the wall, either to recharge the battery or to connect to a modem, when the student was finished, he or she might pull on the cord to bring the ThinkPad closer. Instead of getting up and walking a few feet to pick up the machine, the student would use the cord to drag it a certain distance. This obviously created the possibility of failures we had never anticipated. Who could imagine anyone would treat a computer in this way?

Our colleagues in Raleigh arranged for our engineers to actually speak with the students and ask them how they used their computers. When students arrived into the room to talk, they would often throw their backpacks onto chairs or into corners. The students did not seem to be aware of what they were doing to their computers inside.

Fujino recalls seeing female students twirling their hair while working on ThinkPads, which sometimes resulted in strands of hair working their way into the keyboard. Other students would eat potato chips or other snacks while working or playing on their ThinkPads, resulting in crumbs and pieces of food getting wedged into the keyboard.

Some of the worst abuses happened among younger students. In New Jersey, our engineers saw middle school students playfully hitting each other with their backpacks—which contained their Think-Pads.

Itoh, the most senior woman in my core team, gathered shocking data from four of the schools that tracked breakage rates. Each day, 110 hard disk drives were breaking or being broken at UNC, 70 at Seton Hall, 40 at Union Catholic High School, and 35 at Wake Forest. Each day!

Our engineers kept asking why, why, why? What they discovered is that schools often bought the ThinkPads for students out of their tuition payments. So the students and their families did not directly buy the computers themselves. If they had bought them directly, the students might have been more careful.

The schools also would replace even healthy machines every two years, which increased the sense of impermanence. And the schools had help desks. If a computer broke, students could usually get it fixed or replaced very easily—for free or for very little money. IBM picked up most of the repair tabs because it was impossible for us to determine if a student had abused a system or not. We did not have the forensics capability like that seen on the television show *CSI*. All we could tell was that a ThinkPad was broken. We usually had no choice but to replace the machine because of the three-year warranties we provided.

But the fundamental cultural observation our engineers made was this: American students expected their ThinkPads to work no matter how they treated them. The machines were expensive; therefore, they should work—no matter what. It was the absolute opposite of the Japanese cultural expectation—if something were precious and expensive, we feel we should treat it with greater care. If you examined a ten-year-old ThinkPad in Japan, it would still be in pristine condition.

Not so in American schools. Finally, the message began to sink in: We were not going to be able to change the behavior of American students. We were going to have to make our machines even tougher.

Our Torture Chamber

Our response to the student breakage was to greatly expand the range of tests that we put our machines through. No Japanese would ever think to call our testing facility a "torture chamber"—it was Peter Hortensius and his team in Raleigh who came up with that term. They knew that PC magazines were "torturing" machines to see how durable they were. And when they saw videos of our tests, they applied the term "torture chamber."

One of the earliest tests, dating back to 1992 or so, was the free-fall drop test. This consisted of opening the clamshell case, turning on the operating system, and dropping the system from the height of a desktop.

Then there was the corner drop test. To understand this, I need to explain to you that we thought of the ThinkPads as having Cover A (the top) and Cover B (where the liquid-crystal screen is displayed). Together they formed the top half of the machines. Cover C (where the keyboard and other functional keys are located) and Cover D (the bottom) represented the bottom half. Each of the halves had four corners. Altogether, there were eight corners, and we dropped the machine eight times so that each corner took a direct impact.

The liquid spill test was another test. Starting around 1995, we had designed the ThinkPad so that the critical circuitry was protected from liquids. Water spilled on the keyboard went through the device and drained out without shorting out the key circuits. But we continued testing to see whether milk or soda or other new and different liquids, when spilled on a machine that was powered up, would have any negative impact.

After these tests, we would inspect the machines for damage. The most vulnerable components in those days were the LCD screens and the hard disk drives. The screens could be damaged in many ways, as we had learned from watching American students.

Hard disk drives were vulnerable whenever they were subjected to a shock. Recall that they featured read-write "heads" that were like the needles of a stereo player in the old days. Just as those needles would jump or scratch the record if the machine were bumped, so too could the read-write head malfunction, or worse, break altogether. It had a "parking space" where it could rest when not in use and the disk drive was not as vulnerable in those moments. But if the head were in use and the machine took a hit, data could be lost or, worse, the machine could be ruined.

After all the torture tests, we would then try to make tweaks to our design to better protect these components and work with our suppliers to make improvements on their end as well. It took some of them years to achieve the standards we insisted on.

We made some progress in driving down our defect rate, but IBM's Corporate Quality people were critical of us. We had worked hard to create a culture of understanding and cooperation with the Americans at the PC division's headquarters in Raleigh. We sent Japanese engineers to live there, and they sent Americans to be posted in Yamato. I insisted that my engineers learn English; the Americans were careful to always use the honorific term "san" when they addressed us. We were Naitoh-san and Yonemochi-san to them. We worked very well with Raleigh on design and product development and marketing.

There were cultural differences and disagreements over product specifications, to be sure, but we never allowed our engagement to devolve into a mentality of us versus them. It helped that other nationalities were also represented. We had created an international culture that was highly effective at problem solving. Every weekday morning at seven, North Carolina time, we had a conference call nicknamed the "roll call" where we reviewed the day's issues and what had to be done. We sometimes met on weekends, too. We were part of the same team focused on the same objectives.

But the Corporate Quality people were different. They were part of a separate bureaucracy within IBM and had separate reporting lines. We had no influence with them. The friction with them was an example of how IBM intentionally created conflict within its ranks. Part of the explanation was that we were part of the PC Company, a separate division that IBM set up. The parent company was still overwhelmingly a mainframe computer and server company. Mainframe and server computers rarely broke. They were "up," meaning operational, 99 percent of the time. But we were suffering failure rates in the high teens. We had created a whole new class of products and getting them completely right was going to take more time.

The quality people kept asking us about the next model we were developing. We had to get their approval to ship it, but they kept asking questions such as, "What action have you taken? Can you document that? How can you promise the same issue will not happen on this new machine?" It was not always pleasant, but they succeeded in pushing us to address the root causes of our problems.

We kept on trying. We increased the number of lid closings, assigning people to do thirty thousand openings and closings once per minute. What happened to the hinges? How much torque did they lose? We didn't have equipment to perform these tests, so it was very hard on our people. We gradually started to introduce robots to perform some tests.

It was Sanchez's idea to introduce Tin-Lup Wong to us. Wong was a Hong Kong–born mechanical engineer who had been a professor at Florida Atlantic University before joining IBM in Boca Raton in 1992 and then moving to Raleigh when many of Boca's responsibilities were shifted there in the mid-1990s.

Wong had a PhD in mechanical engineering and was an expert in computer-aided design and computer-aided machining (CAD-CAM). Sanchez asked him to spend three months with us in Yamato to see if he could help.

It took someone like Wong, with his precise mix of cultural and technical skills, to help us make faster progress. He was not Japanese obviously, but he had lived in both Hong Kong and Taiwan and spoke Cantonese and Mandarin plus English. (He carried a British passport because he was born in Hong Kong when it was a British colony.)

Chinese and Japanese cultures are similar enough that Wong knew we had pride in what we had achieved. Partly because of that pride, our organization resisted him at first. But Wong understood that he needed to prove himself. Even though he was from headquarters, he came in with great humility and respect and started mentoring the younger engineers. In addition, he created very solid working relationships with both Ken Yonemochi and myself.

It also helped that Wong had been a professor and knew how to lead people through a logical process of learning. He could convince people to believe what he was saying rather than just telling them to accept what he was saying. That's an important distinction. After he had been with us for about a month, I told everyone on my team that Wong was doing a good job. I supported him.

Technically, Wong was a specialist in creating automated tests, and that proved extremely valuable in managing the problems of the

755C. Intel kept proving that Moore's Law remained valid in part by increasing the amount of power that its new Pentium line of chips consumed and generated. With the previous generation, we were managing an Intel chip that consumed three watts of power. But now it jumped to eight watts, nearly tripling. This was a much higher increase than we ever anticipated. Managing the extra heat generated inside the machine was a problem for us.

We were trying to make the ThinkPads thinner and thinner, about two inches thick at that time. But heat sinks, which were passive devices that used liquid to dissipate heat away from its source, were becoming bigger and heavier to cope with the extra heat being generated by Intel's chips. Fans available at the time were too fragile to survive if a ThinkPad was dropped.

We looked for solutions. We knew, for example, that when NASA built and deployed satellites in space, the "birds" would get incredibly hot on the side facing the sun but would be freezing on the other side. So NASA scientists devised a tube filled with liquid sodium that vaporized when the pipe got hot on one end and drew in cooler liquid from the other end of the pipe. This heat pipe was a very creative approach to dealing with temperature fluctuations.

But we did not think we could apply NASA's approach to our product. It was Fusanobu (or simply "Fu") Nakamura, one of our top mechanical engineers who earned quite a reputation for patterning his inventions on features he found in animals, who made the breakthrough. He saw a variation of the concept in use on a business trip to the United States. He went one day to a store to buy an ice cream cone. The person serving the ice cream had a metal scooping device that had a clever heat pipe in it to very quickly transfer heat from human hands into the tip of the scooper. That warmed the tip and made it easier to scoop up the frozen ice cream. A lightbulb went off in his head.

After he returned to Japan, he found one supplier that could provide a small-diameter heat pipe and started to develop it for the 755C.

We used the heat pipe to direct heat from the CPU to exit on Cover D, the bottom of the machine. Of all of Nakamura's inventions—and he won at least sixty patents over his career—I think this one was the best.

But the problem persisted. The heat was not dissipating fast enough from the bottom of the machine. We had been testing the machines with thermometers and recording the results by hand on paper. But Wong created a system that allowed us to monitor a machine and record its exact temperature characteristics. It involved the use of thermal couples. Each of these couples consisted of two wires of different materials that had been welded together and had something like a pencil tip on one end. We would attach that end to the Think-Pad being tested and the other end to a machine that recorded the temperature. We could do that with thirty to forty couples, sometimes as many as one hundred, to get a precise reading on where heat was going and which parts of the machine were getting hot.

We set standards for permissible levels of heat and began making progress on getting the heat to dissipate in different directions. Gradually, we emerged as the industry leader in thermal design.

Coming In on Little Cat Feet

There were many different kinds of mishaps that could befall a computer, and we tried to harden the ThinkPad against all of them. We knew from our drop tests that there was a different pattern of damage if a machine were dropped from ten inches versus just one. And we knew that a "flat drop" had different characteristics than if one corner of the machine took the impact of falling. The flat-drop scenario, in many ways, was tougher to protect against. In this scenario, a user had the machine open on a desk or a table and realized she was late for another meeting. In her haste, her fingers slip and the machine drops flat from just a few inches. Even a small flat drop could damage the hard drive and cause an unrecoverable data loss.

In 2001, Fu Nakamura set out to conquer that flat-drop challenge and began experimenting with two types of solutions. One was a small air-cushioning pad on Cover D. This shock absorber was filled with air. He learned how to drill a hole of less than 0.1 millimeter into the metal lid on the pads with a laser drill to let air out very slowly. Then after a drop, the metal lid on the pad allowed air to flow back in to return the pad to its original shape and prepare for the next impact.

Nakamura's second invention was a feature called cushioners that came to be known as "cat feet." He learned from many experiments that hard rubber works best to absorb a hard impact such as a ten-inch drop, but that soft rubber is better for a one-inch drop. But obviously, it's impossible to change the hardness of the rubber as a machine falls. The rubber material, of whatever texture, is a constant.

That's why the three-dimensional *shape* of the pads became so important. The pads on the bottoms of cats' feet are textured and variable. Nakamura imitated that shape with a wave-shaped or bumpy rubber material on the cushioner. When a ThinkPad suffers a flat drop from a greater height, the tips of the cushions hit the table first and are compressed. Then the more recessed parts of the cushioner hit. This absorbs the shock. When the ThinkPad is dropped only a short distance, only the tips of the cushion collide with the table, not the full cushion. That absorbs just enough shock to avoid damaging the machine.

We implemented both of Nakamura's inventions into Think-Pad T30 for the first time, but we eventually decided that the cat feet worked best by themselves. We've used them on all subsequent models.

Borrowing from the Auto Industry

But none of the progress we had made was enough. The ThinkPads being used in the corporate world were doing well, but the breakage rates in American schools were still too high. Wong told us that we had to make the American schools our key proving ground. "If we can survive this, we can survive anything," he said.

Having seen students eating potato chips while operating their ThinkPads, Fujino created the dust test to add to our torture tests—a laptop was placed in a chamber and blasted with sand and other long and short fibers to see if the sand or fibers worked their way into the machine. If they did, we would come up with designs to prevent that from happening. We also created a unique test called the "weighted vibration" test as a direct result of seeing how students carried their laptops in backpacks. Wong helped us to develop shock tables and vibration tables to literally torture our machines much like an ancient feudal lord in Japan or Europe might have treated ungrateful peasants. We were merely duplicating conditions seen in American schools.

One other test that Wong helped Fujino develop: the one-hand grab test. We needed it for the T40 because it was much thinner than its predecessor, the T30, which was relatively thick because of the thermal problem. As our machines kept getting thinner, customers could more easily use one hand to hold up an opened computer in a meeting or in a classroom to show people what was displayed on the screen. If the customer were right handed, that meant the thumb on the right hand exerted very intense pressure in a small area of the LCD screen. They also might shake the ThinkPad in ways we had never expected. So we had to learn how to test our screens for intense pressure in small areas while at the same time shaking the machine thousands of times.

The virtue of all the testing was that we knew what it took to break a component. We tortured parts until they broke. We called it "taking them to failure." We knew which supplier's LCD screens were better and which hard drives were better. We had data and we could use it to demonstrate to suppliers that they needed to improve—or else.

We also came up with two new features, both derived from the auto industry—the first was an air bag-like feature that protected the hard drive, and the second was roll cages of the sort that Formula One race cars in Europe rely on.

The marketing guys called it the ThinkPad Airbag. But marketing guys seem to always stretch the truth just a bit. There was no actual bag that exploded inside our machine. Instead, IBM Research suggested that we use one of the accelerometers that the car industry was using to activate air bags but to make it even more sensitive.

It was Susumu Shimotohno, a fellow Japanese and a scientist at IBM's research division, who came up with the idea. The general idea was that if the accelerometer chip on the motherboard detected the possibility of an impact, it would send a special command to the hard drive that would result in the read-write head (like the needle of a stereo) being placed in a safe parking spot, as we called it. If the head were "parked," it could not be damaged and it would not damage the drive.

Shimotohno was smart enough to recognize that there would not be enough time for all that to happen *after* an accident or a fall. So he developed algorithms to predict falls. If there were certain sharp movements of the machine, his algorithms would assume that an accident was imminent and trigger the parking of the read-write head in anticipation that the next shock would be even greater.

I loved the system he created. I told people it reminded me of how a turtle would protect itself in a fall—he tucked his head, neck, legs, and tail inside the shell before he hit the ground. Nothing would get hurt.

The problem was that the system would cost four dollars per machine and that would drive up the overall cost for each ThinkPad. The committee that oversaw the development of each product—which included the general manager, development executives, marketers, and finance people—was worried that it would drive up the price of the ThinkPad or else cut into our profitability. We were undergoing fierce competition in the marketplace and the committee was very skeptical about our idea. It was not a proven technology. No one had done it before. They rejected the idea at first.

It took about a year of argumentation before we were able to pre-

German designer Richard Sapper was responsible for the external design of the ThinkPad and insisted that the TrackPoint in the middle of the keyboard be bright red.

The first ThinkPad, the 700C, captured the imagination of many computer users in part because of its full-color screen. President George H. W. Bush was quick to buy one for his wife, Barbara.

IBM's Raleigh design lab came up with the 701C ThinkPad nicknamed Butterfly because its keyboard stretched beyond the edges of the machine. It had a brief moment of glory in 1995 and became part of the permanent collection of the Museum of Modern Art in New York.

Canadian-born Peter Hortensius served in a variety of roles in connection with the ThinkPad, including vice president of the personal computer division. He was both Naitoh's boss and friend for decades.

Kenshin Yonemochi, a mechanical engineer, was Naitoh's No. 2 in the early 1990s when Naitoh was away from Yamato and played a crucial role in maintaining quality standards for the ThinkPad throughout its history.

David Hill took over as vice president of design for the ThinkPad in 1995 when it was still part of IBM and later made the transition to Lenovo. He is based in Raleigh.

(Left) Kenneth D. Bowersox piloted the first flight carrying a ThinkPad into space and managed to revive it after it appeared to fail. Here he is seen on the International Space Mission in 2002 with a ThinkPad to his left.
(Photo courtesy of NASA)

(Below) NASA official John M. Grunsfeld was instrumental in sending the first ThinkPads into space in 1993 and later became an astronaut. Here he is seen using a ThinkPad on one of his five flights into space.
(Photo courtesy of NASA)

The Yamato lab dispatched one team of engineers in 2004 to study quality problems at American schools. They were horrified to find broken keyboards affixed to the wall of a repair center at Wake Forest University. (Photo courtesy of Kishiko Ito)

The way American students abused their ThinkPads forced IBM to operate repair centers such as this one at the University of North Carolina, seen in 2004. (Photo courtesy of Kishiko Ito)

IBM had to maintain large quantities of spare parts at universities to keep the ThinkPads running, as seen here at Seton Hall University. (Photo courtesy of Kishiko Ito)

In response to the quality problems, engineers at the Yamato lab created tests to simulate conditions seen at American universities. Here is a bump test that simulates how a student might bicycle across campus with a ThinkPad in his or her backpack.

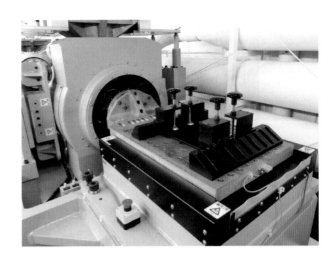

One of the so-called torture tests involved bombarding a ThinkPad in this chamber with dust and other tiny particles to see if they caused the machine to fail.

Engineers had to simulate what would happen if a single point of strong pressure was applied to a ThinkPad.

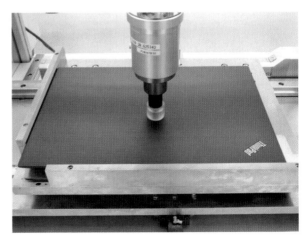

The American management team shown celebrating the ThinkPad's 100 million sales mark in 2015. From left, Peter Hortensius, Fran O'Sullivan, Luis Hernandez, Dilip Bhatia, and David Hill.

Naitoh, at center-left in white shirt, next to General Manager Luis Hernandez, at a management gathering at Hernandez's home in August 2016. The multicultural nature of the ThinkPad team was and remains a major asset.

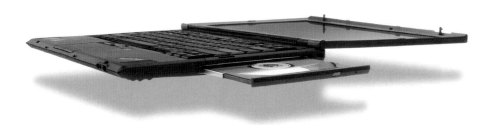

The X300 represented a leap in sophistication when introduced in 2008 but did not sell well because of a world financial and economic crisis. Still, it set the stage for the next generation, the X1.

The ThinkPad Yoga, introduced in 2016, represented the most dramatic shift in the size and shape of the machine, with 360-degree hinges.

The ThinkPad X1 Yoga allowed high school teachers, seen here at Cardinal Gibbons High School in Raleigh, NC, to engage more deeply with students in teaching them how to write. Here, they use NoodleTools, an online research management platform that promotes critical thinking and authentic research.
(Photo courtesy of Cardinal Gibbons High School)

Students at Cardinal Gibbons also use the X1 Yoga in science classes to improve science, technology, engineering, and math (STEM) skills.
(Photo courtesy of Cardinal Gibbons High School)

vail. Wong argued that we should not focus only on the cost of making a ThinkPad but should rather look at the end-to-end cost of selling it and then having to pay out warranty costs. Assuming that we would save on warranty costs, he argued, the real cost of these air bag–like systems would be less than one dollar per machine. Plus, we would be creating greater customer satisfaction—fewer people would lose the data on their hard drives. That line of analysis carried the day.

We announced the feature in October 2003 on two models, the R50 and the T41, but we started rolling it out to more machines in response to the patterns of use we were seeing in American schools. The solution really worked: ever since then, you have never heard about hard disk drives failing.

Wong also played a role in pushing us to embrace the concept of roll cages—metal exoskeletons that protect car drivers in the event of an accident. We resisted it at first because Wong wanted to use light-weight magnesium that was stronger than steel or other metals. He also wanted it to be a unified body—not bolted together from different pieces—to make it stronger. We thought that creating a magnesium unified body would be too expensive and would add too much thick-ness and weight.

But he kept pounding away at us, arguing that any extra expense would be offset by dramatically reduced warranty claims and replace-ment costs. He and some of our engineers in Yamato started examin-ing how roll cages work in race cars.

We resisted for two iterations of the T, the T40 and the T43, but were able to adopt two separate roll cages in the T60, which appeared after Lenovo bought the PC division. One roll cage protected the bot-tom half of the machine, and the other protected the display screen's rear cover, Cover A. We were able to reduce the weight and thickness of other components to make room for the roll cages, so there was no net increase in weight.

We made many other improvements. The usage patterns in Amer-ican schools definitely forced us to take the quality of the ThinkPad to whole new levels. It probably took until 2004 or 2005 for all of our quality processes, manufacturing methods, and products to reach the maturity levels of today. Our failure rates plummeted, and we established a new standard of reliance. Our torture chamber today consists of forty to fifty different tests, some of which are trade secrets. The failure rate of our products is in the mid-single digits. (The exact number is also considered a trade secret.)

But getting there was very difficult. And it put stress on the relationship between ThinkPad and the PC division on one hand, and parent IBM on the other.

Chapter 7:
Crisis Years—2000–2004

My first vacation since returning to Japan in 1988 finally occurred in July 2000. As a matter of policy, IBM asked every employee with twenty-five years of service to take a four-week vacation. This consisted of two weeks of special vacation and two weeks of accumulated annual days off. I should have taken this vacation in 1999 because I had joined IBM in 1974. But I was unable to get away from the ThinkPad business at that time. So I delayed it one year.

As a result of all the traveling I had done, I had a lot of points both from American Airlines and Marriott Hotels, and I used them to take my family to Florida. We went to Disney World, MGM Studios, Sea World, and the NASA Space Center over the course of two weeks. We really enjoyed ourselves. So much time had passed. My daughter was finishing up her university work, and my son was finishing junior high school.

For the other two weeks back in Japan, I helped my son to make a Japanese bow and arrows for his summer vacation homework assignment. We took him to meet with special craftsmen in different regions of Japan who made these bows and arrows. It was one of the only times in his life that his father spent this much time with him. This was a real vacation for me. I was not constantly in touch with anyone at the lab.

All in all, I could have been forgiven for feeling a sense of victory. The ThinkPad had become the number one notebook computer in the world largely on the strength of sales to businesses. For a quarter of a century, I had ridden on the back of the IBM dragon to achieve success.

I didn't realize that it was actually the beginning of a very difficult period, one that would test my own mental health. There were incredibly intense crosswinds. There was the quality problem that I described earlier, but an even bigger problem was that desktop personal computers were gaining such computing power and speed that notebooks were in danger of being left behind. The semiconductors that Intel was producing were using more and more power. From eight watts, they surged to sixteen, then twenty-four, and eventually thirty watts in the 2000–2002 time frame. The more powerful chips were faster, of course, and allowed for better handling of pictures and music. The makers of desktop personal computers could accommodate the additional heat generated by the semiconductors because their machines were much larger than ours.

But these more powerful chips were a much more difficult challenge for us because our machines were so much smaller. To accommodate the new chips, and continue to compete with desktop PCs, we had to accept deteriorating battery life. If our batteries used more power, it only stood to reason that they would not last as long. That irritated customers. We also could not continue to reduce the size of our notebooks because we needed more mass to manage the heat that the semiconductors were producing.

We had traditionally been able to charge a higher price for our notebooks than for a desktop PC, but customers—hit by the recession that occurred after the technology bubble of the late 1990s burst—started to look at us as offering something not as good as a full desktop. So surely we should discount our prices. This economic pressure was very intense for them and therefore for us. "The market changed a lot," Peter Hortensius recalls. "The commercial markets really slowed down because of the financial crisis [that hit in 2001]." I was focused on technology; Hortensius, as vice president and general manager of the ThinkPad unit, was focused on the business.

Sales in terms of number of units continued to increase, but profitability declined because competitors kept discounting their prices. In business terms, the average sales price of our ThinkPads was declining faster than our ability to cut costs. "The challenge was figuring out how to make a dollar," Hortensius explains. "A lot of competitors were doing crazy things in terms of pricing."

IBM had been saying for years that personal computers were a core business, but now I noticed the company was no longer saying that, which worried me. Software and consulting services, plus big computers, seemed to be what the company was emphasizing. They also were selling off many other parts of their Japanese operations. It appeared that they were going to retrench completely from Japan and from manufacturing. The hard drive business, the display business, and the semiconductor business had all been sold. The other labs we had once competed with, at Boca Raton and at Raleigh, had long ago been closed. It felt like we were the last man standing in the face of the inevitable.

Hortensius told me twice that I had to reduce the head count at Yamato. He called this process a "reorganization" or "layoff." I could never use those words. I called them "reassignments" because I found jobs for my people in other parts of IBM's shrinking operation in Japan. Altogether, I moved fifty to sixty people to IBM's consulting division. I never imagined that I would have to do such a thing to my own people.

My Crisis

I did not think it would be the end of the ThinkPad business, but I just didn't know how we were going to fight our way out of the financial mess. At the same time, my mother was dying and she was complaining that I was not spending enough time with her. I was under great pressure on every front.

I developed difficulty in addressing town halls, or "all hands" meetings, of about three hundred employees. This once had been a strong point of mine. I could rally people. But I developed some sort of deep fear of public speaking. I share this story with you not to arouse sympathy but rather to demonstrate how serious the challenges were.

As a general rule, I never went to doctors but this time I needed help. I went to the IBM Medical Center where a doctor referred me to a psychologist. I had two sessions with him and found that these interviews did not help but rather made matters worse. I think many men in the world feel, on some level, that it reflects weakness to go to a doctor. We should just take care of our problems. Those feelings are even more intense for a Japanese male of my generation because of our strong cultural connection to working hard and achieving great results.

The doctor gave me ten tranquilizing pills to help me relax. I carried them in my pocket in case I needed one, but I never did. Just having them in my pocket gave me the boost of confidence I needed. I think, in retrospect, that I developed trouble speaking, at least in part, because I did not have the answers anymore. I felt I was losing control.

One positive outcome of my experience was that I came to better understand how other people at Yamato could fall into similar states of uncertainty. After this, I told my people, especially new hires, that if they ever suffered from mental pressures just too difficult to handle, they should come to see me to talk about them. Not many people actually knocked on my door to discuss their mental anxiety, but I am sure that it helped other people to know that even Naitoh had experienced difficulties. We all were very human.

Innovate or Die

The only thing that was clear during this time was that we would have to focus even harder on innovating and creating new and better

ThinkPads. The motto was "innovate or die" and everyone at Yamato could feel the pressure.

Wireless communications were both a major opportunity and a major challenge. Even though the tech boom had gone bust, the major backbone networks that major telecommunications companies and cable TV companies had been feverishly building with fiber optic cables finally started to reach a point of maturity. Communicating at high speeds on a consistent basis was truly possible. The traffic on fiber optic networks moved literally at the speed of light.

To connect with the high-speed, long-distance networks, we needed short-range wireless techniques. Up until this point, any laptop needed a modem and a telephone landline to connect to the major backbone networks of the Internet. The short-distance lines were so slow that downloading a document took forever, much less a picture or a piece of music. A Wi-Fi system was a router plus radio, and it was typically good for a range of ten to thirty meters. Wi-Fi was useful but there were not enough "hotspots," where the signal was accessible, to completely solve the problem.

Bluetooth became more widely adopted and that was a big game changer. In fact, Bluetooth would not exist if it were not for the ThinkPad. It was General Manager Sanchez, living in Raleigh, who got it started. His backyard neighbor happened to be a chief technology officer for Ericsson, the major Swedish telecommunications company. They started talking over Memorial Day weekend in 1998, each probably with a beer in hand. Sanchez explained that he and his team essentially wanted to build a phone into the ThinkPad so that it could communicate at any time and from anywhere. But that was too expensive and involved too much bulk.

The Ericsson executive explained that his company had short-wave radio technology that was very fast. It could replace any physical wires and rapidly transfer voice and data. (The name "Bluetooth" referred to a medieval Viking king, Harald, who had a dead tooth in his mouth that appeared to be blue in color.)

Sanchez suggested that they try to make it an industry standard. That often happens in the technology world. Before a new idea can truly take off, it has to be established as something that many different companies will use. Ericsson wanted Intel to be part of the Bluetooth consortium, and soon Toshiba and Nokia were members. The first version of Bluetooth was released in 1999, and it rapidly expanded and was improved. This meant it was now possible to connect any device in a room to a short-range wireless system that in turn connected with a long-distance, high-speed network.

The world was now accessible without any wires. It was the beginning of today's era in which personal electronic devices could be truly free—you did not need to plug them into the wall for power and you did not need a modem line to communicate. Digital cameras also became increasingly popular and affordable at about this time. People no longer had to buy silver nitrate film and then—once they took their pictures—take the film into a shop and wait days for it to be developed, a major shift in photographic habits. Music was also being digitized—a company called Napster shook up the music world by allowing peer-to-peer sharing of music. But it was still early for video.

We had to make major modifications to our ThinkPads in response to these new trends. The first problem was that Cover A, or the top of the machine, was made out of carbon fibers that blocked the radio signals that would have to be transmitted to connect the machine to the Internet. We loved our carbon bodies because they were lightweight and very strong. But we had to find a different material that would allow the signals to be transmitted wirelessly.

Ken Yonemochi by now had moved from mechanical engineering to product assurance, so it was Hiroaki Agata and his mechanical engineering colleagues who introduced the use of glass fibers. The glass fibers allowed the radio signals from inside the machine to be transmitted through them. We decided to replace a small part of the carbon fiber on Cover A with the glass to create enough room for those signals to get through.

The very real problem was how to combine glass and carbon into one smooth surface that appeared to have the same feeling to the touch and displayed the same color. There was a joint between the carbon and glass but we had to make it seamless and smooth. We had to learn how to sand, polish, and paint the two different materials to be the same color.

A related problem was with the antennas that would transmit the wireless signal. We started with two Wi-Fi antennas built inside the top half of the computer and positioned them so that they could transmit their signals through the glass fiber. We discovered that if we put the antennas in the bottom half of the computer near the keyboard, they did not work as well. Depending on precisely where we placed the antennas, there also could be a problem of degradation of their signals due to electromagnetic interference from human hands. We could not place an antenna too close to where a user's hands typed or scrolled.

Because it was so important to maintain seamless connections with the Internet, we had to embed several antennas to support Wi-Fi, Bluetooth, and wide-area networks (cellular telephone networks). There was a tremendous rush throughout the industry to come up with the smallest and most powerful antennas for all these systems. Our engineers tried to drive down the size of antennas to just a few millimeters in surface area at the same time that they raised their decibel level so that they could more easily connect with access points located in an office. We performed endless experimentation with this, and I was personally involved in driving my engineers to get it right.

Imagine you are in a meeting room with a big conference table in the middle. You put your ThinkPad on the table and open it up, placing the lid at a certain angle. The signals would be transmitting from the back of Cover A in one direction. But what if there were no access points in that direction? The computer could not connect. One prominent computer magazine got hold of one of our machines and tested its connectivity. It did not give us a very good rating.

I kept up the pressure by showing my engineers a picture of an office and challenging them to demonstrate how a wireless laptop could perform in that environment. What if we had two antennas? What if we had four? Which way should they all be pointed? And very importantly, how could we enable our machine and its software systems to smoothly hand off and recover the flow of communications from one access point to another, assuming that perhaps the owner of the laptop might be moving? We needed smooth switching.

If you pick up your ThinkPad today and try to find any evidence of what we struggled with, you will be disappointed. All this work was, and remains, completely invisible to the end user. Altogether there are seven different types of antennas in today's machines. Over time, we provided users with an enormous new capability—the power to connect at high speeds through thin air to the Internet. It was a kind of magic. And we were the first company to allow it to happen in a notebook.

The Owl Wing Fan

Fu Nakamura, the heat pipe and cat's feet engineer, kept attacking our thermal problems. He had been working on the problem for ten years when he came across the concept of an owl wing in 2003. His colleagues liked to tease him about his crazy ideas, but he kept finding ways to make them work.

The reason we had not installed fans in our ThinkPads for the first few years was that they were too fragile. But over time, a new robust shaft-bearing technology allowed us to start installing fans in a way that made them much sturdier. They were, however, big, and they made noise. We had very strict guidelines about how much noise we wanted our notebooks to emit. Fans also consumed a great deal of power, which put pressure on battery life.

As Nakamura attended meetings and conferences of noise reduction engineers and heating engineers in search for answers, he heard the story of the owl. An owl can fly very quietly. It can swoop down and catch small animals such as mice before they know what hit them. Those animals were very sensitive to any noise, so the owl had to make virtually no noise.

Nakamura did not experiment on any living owls and did not watch a special program on the Discovery Channel. Instead, he heard about owls' wings from other engineers who had worked on Japan's bullet trains, the *shinkansen*. These trains routinely reach speeds of 200 miles per hour as they tear through the Japanese countryside. They get their power from an electricity line above the train, and they have metal arms on top of the train, on both sides, that capture the power from the electricity line. They float on top of the line because there can be no fixed connection, of course, if the train is speeding along.

The problem the bullet train engineers had to solve was the noise that was created by the metal arms being in constant contact with the electricity line. That friction created a kind of screeching sound that many people hated.

In order to cancel or decrease the acoustic noise, the bullet train engineers drew on some of the properties of an owl wing. Imitating how the feathers are aligned on the bird's wing, they put bumps on the base of the metal arms that gave them a different air-flow profile as they moved through the air. Those bumps reduced the vortices of air that created noise. These vortices had been discovered by a Hungarian engineer, Theodore von Karman, many years earlier.

The *shinkansen* engineers were imitating the owl's wing on objects that were moving straight. Nakamura's challenge was to use the concept on the blades of a fan that were rotating in a circular motion.

At first, it did not work. So he started doing more experiments, including using computer simulations to study various shapes of owl wings. He learned to make a small cutout in each blade of a fan, which canceled the vortex effect. Rather than one big noisy vortex, there was a series of smaller vortexes that did not create nearly as much noise.

Nakamura visited me many times to talk about his work. I was excited because I could see how valuable this could be. He started showing me data about the fan blades, but I didn't think the data was convincing. So I asked him to keep working on the problem.

I told Fran O'Sullivan, then the general manager of the PC division, about Nakamura's work. She said she had two owls in the backyard of her house and that "they are noisy birds." She didn't accept our argument. She said owl wings made a great deal of noise.

I went back and told Nakamura, but he figured out the secret. At certain frequencies, our fan blades might make too much noise just as an owl might make noise when it first takes off. But at other frequencies, our fan blades would be virtually silent, just as the owl's wings are as it dives for its prey. Nakamura argued that an owl's wings were even quieter than those of an eagle as it made its swooping assault.

That was it! It was all about the frequency of the blade rotations.

In addition to that breakthrough, we learned how to improve the airflow through the entire cooling system, from the point of air intake all the way through. Taken together, these breakthroughs allowed us to reduce the size of a fan unit by 20 percent and make it viable to be built into a ThinkPad.

Nakamura became known as Mr. Owl Wing. His invention changed the very nature of our ThinkPad. We could finally and decisively beat the heat problem, and the fan was small enough that we could continue to drive down the thickness of the ThinkPad. The first ThinkPad to incorporate his breakthrough was the T60 announced in 2005. "Everybody used to laugh at me," Nakamura recalls triumphantly. "Now they want to hear how we did it."

Handwriting on the Wall

In October 2003, ThinkPad management asked me to move to PC division headquarters in North Carolina and become chief technology officer for the whole division, not just for ThinkPads. It was a promotion and hard to turn down, but there was more to it: key executives knew that IBM was negotiating to sell the division to Lenovo.

One of them was Peter Hortensius. I didn't realize at the time what his strategy was. He was trying to keep me close to him and engaged in trying to invent the future. He later explained it this way: "We knew that if we could keep Naitoh-san's head in the game, we could keep the Yamato lab's head in the game. Yuanqing [the Lenovo CEO] was very clear with me about that right away." (Yang is his family name. Yuanqing is his given name. It is pronounced Ya-wan-ching. The "q" in Chinese is a "ch" sound in English. The fact that Hortensius could call him by his given name suggested a high level of familiarity.)

In other words, Lenovo's top management recognized the value of the Yamato lab and was telling IBM, as it negotiated the deal, to make special efforts to retain our loyalty. Hortensius figured that if I was at headquarters, and eventually brought in to the negotiating process, I could understand what was happening and perhaps help shape it. People at the Yamato lab still looked to me as their leader, even if others had assumed the daily functions I had been performing. Whatever personal doubts I had been having privately, I was still seen by many others as the key strategist.

The handwriting was on the wall by the time my wife and I arrived in Raleigh. Hortensius thinks senior management had been working on selling the PC division since 2001. We were clearly not going to be successful as part of IBM. Here is how Lou Gerstner described the situation in his 2002 book, *Who Says Elephants Can't Dance? Inside IBM's Historic Turnaround*: "Perhaps the most difficult part of the business that needed to be overhauled during my tenure at IBM was the PC segment of our portfolio. Over the course of nearly fifteen years, IBM had made little or no money from PCs. We sold tens of billions of dollars' worth of PCs during that time. We'd won awards for technical achievement and ergonomic design (especially in our ThinkPad line of mobile computers). But at the end of the day it had been a relatively unprofitable activity. There were times when we lost money on every PC we sold."

Gerstner said the biggest reason we had never been able to make enough money was that Intel and Microsoft controlled the key hardware and software elements of our machines, which they did. They had dominant positions in their respective fields and could demand higher prices than their rivals might have. That's why they made money and we were squeezed. Gerstner's analysis set the stage for the inevitable.

Chapter 8:
Cracking the Mysteries of the Nile—and the Planet

When adventurer and explorer Pasquale Scaturro decided in 2003 to lead the first team to descend the entire length of the Blue Nile and the Nile Rivers, he included three different notebooks from different manufacturers on his equipment list. One of them was a ThinkPad. It was not the lightest notebook he could find, but he trusted the IBM brand name. He felt the machine would not break as he descended from the almost ten-thousand-foot highlands of Ethiopia, through Sudan, to the Mediterranean Sea at Alexandria, Egypt.

Scaturro, born in Hollywood to a Sicilian father and American mother, needed a secure way to be able to file reports and journal entries to IMAX, the company that was paying for the journey to make a movie called *Mystery of the Nile*. The mystery revolved around what was the real source of the mighty Nile—the White Nile or the Blue Nile tributaries.

Many explorers dating back to the ancient Greek and Roman times had attempted to travel up the river from Egypt, but had never made it to the source of the Blue Nile, which provided the lion's share of the water in the Nile. The White Nile was much easier to explore and conquer because it was a much gentler ride. Many explorers mistakenly identified it as the real source of the Nile.

Other explorers such as the Portuguese and British had hiked to various points along the Blue Nile and traveled segments of the river, but none of them made the entire descent to the Mediterranean "from source to sea" as Scaturro described it. The hazards were too extreme. Rivers have different classifications depending on how difficult they are to canoe or kayak. A perfectly smooth river is Class I. But segments of the Blue Nile were classified as Class VI, the highest level of difficulty, which basically meant they were impossible to navigate. They had to be portaged. Then when the river descends from the highlands and calms down somewhat, the major hazards become crocodiles and bandits called *shiftas*. Many would-be explorers over the years died or disappeared.

Scaturro's journey began in late 2003. He does not remember exactly which ThinkPad model he took, but judging from the timing, it must have been one from the XTRA generation. Of the three computers, the ThinkPad emerged as his primary workhorse. "I used the ThinkPad extensively," he says, speaking from his home in the Namibian desert, the least-populated part of the second-least-populated country on Earth.

"I had to send back dispatches," he explains. "I would do that every third or fourth day." Altogether the journey of 3,500 miles lasted six months.

He built a solar charger, consisting of small solar panels connected to a little battery to charge the ThinkPad. He could then hook the computer up to an Inmarsat satellite telephone system. Communications speed was somewhat slow, but it worked. He was way ahead of the rest of the world in leveraging possible ways of using the ThinkPad.

Whenever conditions permitted, at camp in the evening, Scaturro would set up a small portable table and write journal entries on the ThinkPad. One journal entry from day nineteen, while he was in Ethiopia, is particularly vivid: "In reality, man can only do so much to the earth. If everyone in the highlands died tomorrow, all of Ethiopia would probably revert back to wilderness quickly. In the scheme of things, man is pretty small and insignificant."

Thanks to his clever communications setup, he was able to send out eighty thousand words over the course of his journey, which became the foundation for a book by the same name as the movie.

Just as it was surprising to some that the ThinkPad survived in space, so too was Scaturro impressed that it survived so many dunks into the river. It was carried in a watertight Pelican case aboard one of two large yellow rafts, but it was still handled very roughly. It survived temperatures of 130 degrees Fahrenheit in the equatorial part of the journey, where the river passes through blistering desert sands. It also survived an encounter with Egyptian policemen who arrested the crew after they crossed the Egyptian border without official governmental permission. "It was a real tough computer," Scaturro says. "The keys didn't stick. The operating system was more stable than other machines. It just didn't break."

The ThinkPad was not the only factor, or even the decisive factor, in allowing Scaturro to undertake his journey. But people like him embraced the ThinkPad and used it to explore the literal ends of the earth. In 2006, Lenovo partnered with the Explorers Club to provide ThinkPad X60 ultraportables to five environmentalists who wished to study biodiversity in rain forest canopies. They each weighed three pounds (1.36 kilograms). "Collecting forest data in diverse environments, often high above the ground, is challenging for any scientist," Dr. Margaret Lowman, director of environmental initiatives at the New College of Florida, said at the time. "The forest canopy is our office, and our notebook PCs need to be lightweight and reliable enough to withstand these environmental conditions."

Several climbers took the ThinkPad up Mount Everest starting, to the best of our knowledge, in 2002. It was a Japanese college student, Atsushi Yamada, who approached us and explained that he wanted to be the first person to take a computer to the top of the world's tallest mountain. He wanted to establish a world first in that regard while at the same time he was competing to become the youngest person to climb the tallest mountains on the Earth's seven continents. Everest would be the last for him. He wanted to be able to say that he operated the machine on the top of Everest.

We knew that one key vulnerability of our machines was the hard disk drive. As you may recall, the read-write head floated just above the disk. This was an initial worry when we contemplated NASA bringing ThinkPads into space. That did not turn out to be a problem because the astronauts operated the machines in pressurized cabins and in the pressurized space station.

But now Yamada was proposing to operate an X32 ThinkPad at more than 29,000 feet of elevation, where the air pressure is very low. The read-write head would certainly drop and touch the disk, causing the system to crash.

Other challenges were that liquid existed in both the batteries and the screen. If exposed to extremely low temperatures that prevailed at the top of Everest (an average of more than 30 degrees below zero Fahrenheit), those liquids could freeze, causing the machine to fail. In some ways, these conditions were more extreme than anything the ThinkPad encountered in space.

To get around these problems, we made special disk drives with shock absorbers and a frame of titanium alloys for Yamada. He carried the battery pack inside his jacket to keep it warm on his ascent. He made it to the top, got the battery out, and inserted it into his ThinkPad. It booted up successfully, but the battery lasted for only a minute in the harsh conditions. It was long enough, however, for him to establish his claim that he had done what no one else had done before him.

Yamada was not able to communicate with the world in real time. Battery life was one problem but satellite communications systems also had not been fully developed.

As I will explain in the next chapter, we started making progress on the hard disk drive problem by adopting solid-state hard drives that were not as vulnerable to gravitational forces and air pressures. More climbers started bringing these ThinkPads at least to their base camps.

In May 2012, climber and guide Eric Remza, a thirty-seven-year-old American, took a ThinkPad X220 equipped with a solid-state drive to his base camp at Mount Everest. His sister worked for Lenovo in the United States, which is one reason he was able to obtain the very latest model. Remza was going to serve as one of several guides for an expedition organized by International Mountain Guides, based near Seattle, Washington. He wanted to be able to stay in touch with the rest of the world while away. In an email interview with Lenovo from his base camp at about 17,800 feet, he explained that temperatures could get so cold that the contents of his pack, such as water bottles and sunscreen, would freeze. His ThinkPad encountered those same temperatures and the same extreme variations, depending on whether the machine was in the sunshine or not. "I do need to take some care with keeping it protected, but overall the solid-state drive is running like a top and starts up in the cold with no issues," he said. He was also able to transmit photographs.

The solid-state drive seemed to conquer the low air pressure problem. "If your electronics are relying on a cushion of air to process, then they will surely eventually 'crash' when subjected to the differential air pressures at these elevations," Remza added. On a previous assault on Everest in 2007, starting in Tibet, he and his party brought laptops to their advance base camp at 21,500 feet and lost several of them because their hard disk drives relied on air cushions to function.

The base camp was well established and had a communications tent that used solar chargers and batteries to recharge any personal devices climbers brought with them. At first, Remza used Ncell air cards to communicate. These were sold by a private mobile telephone company operating in Nepal, where the primary approaches to Mount

Everest are located, but they didn't work well because so many climbers were trying to use them at the same time. The circuits, in effect, jammed. He ended up relying on a friend's Thuraya satellite phone to send out his dispatches. "For me, staying connected in the mountains via the right computer is as important as having the right Sherpa staff and technical equipment," he said.

One other reason the ThinkPad was valuable to him and his expedition was that it allowed him to anticipate weather conditions. The way climbers stage their climb is that they reach their base camp and start what they call "acclimatization rotations." Each day, they make climbs to reach a "high point touch" and then return to base camp. They might make three of these rehearsal climbs. Then they rest for as long as a week at base camp.

The final assault takes four days and it's critical that weather conditions on the fourth day be just right. "You want ideal conditions with low winds and good visibility and low precipitation," Remza later explained. "It's all about waiting for the right time."

So, using the ThinkPad and other devices, Remza and fellow guides could communicate with meteorologists who have experience with weather patterns in mountain ranges. The climbers paid for the best weather insights. "If you have twenty to twenty-five people on the expedition, you need that to safeguard their safety getting up and down the mountain," he explains.

Remza did not take the ThinkPad with him on the final four-day assault in 2012. There would have been no way to recharge the device and satellite coverage remained spotty. Plus, even if the machine weighed just three or four pounds, it was extra weight. And on such a grueling climb, every ounce counted.

To our knowledge, no one has yet succeeded in climbing to the top of Everest and transmitting live from a ThinkPad. In 2013, a British explorer used a smartphone to transmit live video from the summit to the BBC—but it was not a ThinkPad. So the ultimate challenge of using a ThinkPad to communicate from the top of Everest remains to be conquered.

As you can see, the ThinkPad has encouraged scientists and explorers to test their personal limits and the limits of their equipment to explore untamed rivers, to catalog the species living in rain forest canopies, and to climb the most difficult mountains. In so doing, they have changed the nature of exploration. It used to be that explorers and adventurers disappeared for weeks or months and then, if they survived, they came home and told their stories. Now, however, explorers can communicate with audiences and make their adventures more of a participatory experience. We think we helped fire all their imaginations.

Chapter 9:
Crossing the River to
Lenovo — 2005–2008

While living in North Carolina with my wife in the summer of 2004 and working as chief technology officer for the PC division, I was "disclosed." That meant I was brought in on a secret—IBM was negotiating with a few companies to sell the PC unit, including the ThinkPad unit and, of course, the Yamato lab. Hortensius broke the news to me.

I did not jump out of my chair. I was not surprised. The trend line had been headed in that direction for some time. I was more curious than anything else. I was not the head of the Yamato ThinkPad development team. People there still had deep loyalty to me and I to them, but another executive was head of the lab. For now, I was on the sidelines.

Publicly, the news broke on December 8, 2004, that IBM had agreed to sell its PC division to Legend, as it was called at the time. It was the first time that many of us had any idea that a Chinese company was even in the running to acquire us. None of us had ever heard of Legend, which was in the process of renaming itself Lenovo.

A town hall meeting with all hands in Yamato was called to discuss the sale. I could not monitor the event from North Carolina on my computer because online meeting technology had not advanced sufficiently. But I had numerous phone calls and email exchanges with people in Yamato.

It clearly was not going well. People were panicking about their futures. We were worried that engineers and others would start resigning en masse. I went to Hortensius and said, "Yamato is no longer my territory, but considering the situation, I think I need to go back."

He agreed emphatically and off I went with my wife. I didn't know how long I would be gone, so we went together.

When I got back to Yamato, I started with a whirl of small meetings and then a town hall meeting. People said they felt it could not be real. I empathized. "I agree," I said. "This doesn't feel real."

Many said they selected IBM to be the company where they would work their whole lives. I empathized. I also had chosen IBM for my life's work, and I was an IBM Fellow, a huge honor that now seemed diminished.

But after empathizing with them, I told people that the PC business no longer makes sense for IBM. The hard drive, display, and semiconductor businesses are all gone. We have to go, too. If we don't take this chance, IBM might just shut us down. We created the ThinkPad brand. We will take the ThinkPad and survive.

I was speaking persuasively. My previous speaking problems had disappeared. I'll never fully understand what clicked inside my head. Perhaps now the situation was much clearer than it had been before. Once again, I had conviction.

I said, "We have to cross the river." This is a potent symbol in Japanese culture. If we all hold one another's hands or arms and cross together, no one will slip and fall in and die. I told them it was no longer tenable to stay on this side of the river. And Lenovo was the best bridge. "Trust me," I said.

One of my goals, of course, was to prevent valuable people from leaving because of all the uncertainties. Our key engineers said they had never heard of Legend or Lenovo, and loved IBM too much to accept this acquisition. Being part of big IBM gave them a sense of security. But I told them that IBM no longer wanted them. I asked them, "What do you love most? The ThinkPad or IBM?" Of course, their answer was the ThinkPad. I persuaded many other people by asking them, "What are you really looking for in life?"

The national cultural issues could not be escaped. Newspaper stories and television reports suggested that relations between the Chinese and Japanese governments had not been pleasant. Some Japanese feared the size of China—they have more than ten times as many people as we do. Others worried that to settle old scores, the Chinese would take our jobs and our technology and move them all to China. The pressure on employees was intense. Their extended families were asking them how they could justify going to work for such an unknown foreign entity. As I said, IBM was not considered foreign. We saw it as Japanese.

In strictly business terms, I argued that it made more sense for Lenovo to own us than for IBM because Lenovo was a personal computer company. They wanted to move into mobile device fields such as telephones, but first and foremost they made personal computers. We were central to their mission in a way that we were never going to be at IBM. Big Blue had many different types of businesses—software, consulting, mainframes, servers, and others. Lenovo's primary focus was on personal computers. It would be much easier inside Lenovo to understand how each employee could affect the company's overall results. We could have a much stronger presence and sense of purpose.

I stayed in Yamato until February making these points over and over again. "Trust me," I kept repeating.

Altogether, we lost only fifteen out of three hundred full-time employees because of the Lenovo acquisition (and we had another two hundred contract employees). We had preserved the heart of Yamato.

Foggy in Beijing

I returned to Japan full time in March 2005, just before the Lenovo acquisition was scheduled to officially close. As of May 1, when the deal closed, I became Lenovo's vice president of Notebook PC development, with both the ThinkPad team in Japan and the IdeaPad team in China reporting to me.

Chairman Yang Yuanqing had been communicating with me and invited me to go to Beijing where he promised to hold a dinner in honor of a handful of us from Yamato and Raleigh. He told me to call him by his given name, not his family name. He called me Naitoh-san, using the honorific term that means something like "mister" in English. So very early in the relationship, he took a very modest position, almost as if he were looking up to me. He was at least a decade younger than me so that was also part of the dynamic between us. In East Asia, people still defer to their elders, which does not happen as often in the United States.

Soon after the deal closed that May, my team planned to go to Beijing to start three days of technical discussions with our counterparts. This was the first step in bringing the research and development efforts of the two companies together. Yang scheduled the dinner.

When we started getting ready to meet Chinese engineers we would work with, we speculated privately that they would turn out to be very different from us. But almost immediately, I realized that engineers speak the same language everywhere. There was no difference between us just because of nationality. I sometimes see a bigger cultural gap between Japanese people from different regions or walks of life than between two engineers, one of whom happens to be Japanese and the other Chinese.

At the last minute, Yang had to change plans and dine with a customer instead of us. We all understand how important customers are. That is universal.

To make up for the missed dinner, after a morning of technical meetings, Yang scheduled a lunch at a restaurant in the Crown Plaza North Hotel. Because he had initially promised a dinner, he wanted to make the luncheon as special and memorable as possible. So he organized a Chinese banquet with *baijiu*, or white liquor. This was similar to the *maotai* that President Nixon drank on his first visit to China. Both are strong and fiery. In both China and Japan, drinking is an ancient way of creating bonds, of getting a sense of who a person really is behind the layers of formality.

Yang handled his liquor well, but just in case he was challenged by someone who could hold their liquor better, he brought another executive who had a military bearing and was obviously an experienced drinker. Yang kept toasting to us and saying *ganbei*, or "dry bottom," the equivalent of "bottoms up" in English. It was very similar to what we said in Japanese, *kanpai*. It is one of the few expressions that sound similar in both languages. After he finished a round of toasts, his companion would launch another.

In the etiquette of drinking, it was a loss of face to not accept a toast. However, I did not accept every drinking challenge. So I remember Yang saying that the ThinkPad group was "the pearl in the company's crown." In English, you would say we were "the crown jewel." Linguistic subtleties aside, what he meant was that our skills and technologies were important to the future of Lenovo. This acknowledgment was important.

I don't remember all the details of the lunch, which lasted for two hours. Kishiko Itoh, one of my engineers who was traveling with me, remembers more than I do, so I must have gotten a little bit tipsy. I think many of the Japanese and Americans in our group might have felt that way. When we went back for more technical meetings that afternoon, I had to shake it off. I remember asking, "What are we talking about here?"

Yang later came to Yamato and gave a speech to all hands. He was not hiding anything. He was very honest. He wanted to communicate how he felt about us. That evening, a group of us went to a restaurant with him and had *shabu shabu*, which is the kind of Japanese cuisine where you cook thinly sliced beef and other foods in a cooking pot. We sat on *tatami* mats in the best Japanese tradition. It was not a karaoke place so it did not have microphones, but Yang and a colleague sang a song in the Japanese language for us. It was called "Springtime in the North." I learned that it was one of the most popular Japanese songs in China and many people can sing it in Japanese. Yang was demonstrating deep respect for us and our culture.

As my father would have done, I started to try to articulate a different vision of Japan's relationship with China to all my people. Aside from periods of hostility, there also have been periods of great cooperation and learning. About 1,200 years ago, for example, the Japanese sent delegation after delegation to the Chinese capital at Chang-an (now called Xi'an) and engaged in the wholesale importing of Chinese architecture, written characters, brushes, tofu, and other important contributions.

Our ancient capital of Kyoto, which attracts tourists from all over the world, was copied on the basis of what we saw in China. We even incorporated the Chinese concept of *feng shui*, which literally means "wind water" but, of course, is a large body of belief about how buildings should be built and rooms should be furnished to allow the sacred life force called *chi* to circulate freely. As a result, for example, we decided that our new Imperial Palace should face south, an auspicious direction.

I could also see that Lenovo was rapidly internationalizing itself. It adopted English as the official operating language. All its top people and key technical talent were committed to learning English. So we Japanese would not have to learn Chinese. We could communicate in English and would not have to learn a third language. That was important to us.

Lenovo was clearly adopting the management style, salary policies, and other practices of IBM and other Western multinationals, and I thought we would be able to understand how decisions were made. There would be greater transparency than we might have expected from a traditional Chinese enterprise or company, where a small group of people or just one chief executive typically makes decisions in private. Lenovo's key decision-making body was the ten-person Lenovo Executive Committee, which was headed by Yang but included nationals of five or six other countries. The executive committee stayed in motion. One month, they were in Europe. Another month they were in North America. Wherever the executive committee was, that was considered headquarters. I didn't feel that we would always have to go to Beijing to seek decisions.

It also helped that Lenovo decided to maintain Raleigh as one of two official headquarters, along with Beijing. It was serious about continuing to run many parts of the business from Raleigh. That was a very sophisticated decision. Not many big, established companies from Europe, the United States, or Japan would even try an arrangement like that.

We had indeed crossed the river.

The First Glimmer of the Yoga

As I mentioned, I was now head of ThinkPad development and at the same time I was running a notebook lab in China. The ThinkPad was still aimed primarily at the business and education markets, but the lab in China was working on a line of products called the IdeaPad aimed primarily at consumers.

Coordinating these two labs was a very delicate task not just because of different nationalities and cultures but also because every technology development organization has its own terminology and its own processes. If the integration was not done well, or not done at all, it could have created chaos in our ability to develop new products.

But I could see that if it went well, we could unleash new competitive energy from the much younger Chinese engineers. We could tap new ideas. On one of my first visits to Beijing, I was invited to an industrial design lab and saw a mock-up of a new kind of notebook computer sitting on the desk of Yao Yingjia, who would become famous for designing the torch used at the 2008 Beijing Olympics. The mock-up was very primitive and was made of wood. It looked like two sides of a small tent. It was just a toy.

"What's this?" I asked him.

"It's my innovation," he answered. "It is just an idea."

This was the first time I saw the concept that would emerge as the Yoga four-way convertible ultrabook and would ultimately enter the ThinkPad line of products. It had a 360-degree hinge that allowed the top and bottom of the computer to be moved into four different configurations—for use as a notebook, for use as a tablet, for use in watching movies in the "tent" mode, and for use, say, in a kitchen looking at a recipe. This was a very important seed of an idea that would blossom.

We now had three centers of innovation—Japan, China, and the United States. Japan and China were where the hardware work was being done; the States took the lead in software, marketing and branding, and quality control. We called this our innovation triangle.

There were enormous skills gaps between the Chinese engineers and us. We had been functioning as a successful world-class organization for at least fifteen years. Our engineers were beginning to show some gray hair. Many of us were in our fifties.

There were very few older Chinese engineers. Atsushi Kumaki, one of my best software engineers who had been with me since the days of the L40 SX, spent four years in China and got to know many of these younger engineers who were in their twenties and thirties.

Partly as a result of the age and experience gap, there were very different attitudes. We from Yamato had seen everything go wrong with a computer that could go wrong and we always anticipated issues and problems. When confronting a new idea, our basic default position was to explain why it couldn't work.

But the Chinese were aggressive. Why not try it? That was their attitude. "The Chinese are very free in their thinking. They like out-of-the-box thinking," Kumaki explains. "We Japanese guys with twenty or thirty years of experience will say, 'This doesn't work. We know it from experience.' We are stopped by ourselves. Experience sometimes blocks us from doing something."

In terms of the style of innovation, "we had an accumulation of step-by-step knowledge about how to make very high-quality things," Kumaki says. "But we Japanese are not good at seeing the big picture first. We are good at doing it from the ground up. But the Chinese see the big picture, the big goal, and try to achieve it even if they ultimately cannot achieve it. They jump in."

This can-do attitude was refreshing to us even if the Chinese were still far behind in their overall technical capability.

I invited many engineers from Beijing, Shanghai, and the southern manufacturing city of Shenzhen (where our machines were assembled) to come to Yamato to see how we developed ThinkPads, and I sent members of my team like Kumaki to spend longer periods in China. I also sent mechanical and wireless engineers. They tried to address specific technical problems the Chinese teams were facing.

Throughout this period, I was very positive about transferring knowledge. Some of my people, reflecting the traditional distrust between the two cultures, asked me, "Can we tell them everything?"

My answer was, "Many people think that to keep a strong position you need to hide something. That's wrong. If you feel you will be empty after you tell them everything, you need to go out and learn more."

Project Kodachi

At the same time that I was managing the integration of these research efforts, I was aware that we were facing overwhelming pressure from customers seeking reassurance that the soul of the Think-Pad—our commitment to quality, reliability, and innovation—would not change because Lenovo had acquired us. "Every customer we met with asked us that question, 'What's going to happen to the quality?'" recalls marketing executive Jerry Paradise, who was then technical assistant to PC division general manager Fran O'Sullivan. "There was a presumption that we would follow in the way of many Chinese products. We would replace our engineers with low-cost engineers and the quality would go down."

Chairman Yang also could feel the pressure and clearly wanted us to prove something. "We want to send the message that if there's a company in the industry that can continuously develop the most inventive and best-quality products with efficiency, it will be Lenovo," he was quoted as saying.

Everybody inside the company wanted to make a statement. We wanted to create something of such high quality that the skeptics would be silenced. Design chief David Hill was instrumental in managing the decision-making process to do that. This project was code-named Kodachi, referring to the smaller of two swords that a samurai warrior traditionally carried.

We started the project in June 2006, roughly a year after the acquisition closed. Hill wanted the machine to be less than one inch thick with a ten-inch screen, yet he wanted it to have a full-sized keyboard. He was thus returning to the "butterfly" concept we had tried in 1995. Hill worked with us in Yamato and with other design colleagues in China as well as designer Richard Sapper, who had survived the transition from IBM to Lenovo. Every day, we sent drawings from Yamato showing how various components and electrical parts might fit together.

By late September, it was clear to us that thirteen-inch screens were becoming popular on notebooks for watching movies. Video use on all manner of handheld devices and computers was exploding partly because the underlying communications networks had been upgraded and could carry higher volumes of data. Ten-inch screens were clearly out of date. Hill conceded the point and we dropped the idea of having a keyboard hanging out over the edges of the computer's body.

To make a clear statement about our intentions to continue to lead the industry, we agreed to use three key technologies for the first time. The first was a solid-state disk drive using a flash memory technology called NAND. (This is what the climbers began using to climb Mount Everest.) Replacing the hard disk drive that had historically caused us such headaches, these new drives used semiconductors, not a manual read-write function, to preserve data. It was high-speed and relatively expensive, but it was physically smaller than a mechanical disk drive. That would allow us to drive down the size of Kodachi.

The second was light-emitting diode (LED) backlighting on the screen, which would improve movie viewing. Before you can see anything on a computer screen very clearly, there needs to be a source of light behind the screen. We had been using fluorescent light, but an LED light was more efficient and thinner. This was another big change in the architecture of the ThinkPad, but it was again aimed at creating an ultrathin machine.

The third was an optical DVD drive from Panasonic. This was another element intended to facilitate the use of video. Drives started out being one inch high in desktop computers and soon were reduced to half an inch, or twelve and a half millimeters. We demanded that it be reduced even further. Panasonic reduced it to nine millimeters. It still was not good enough, we insisted. They finally reduced it to seven millimeters. We were the first to use a seven-millimeter optical disk drive (also the last, because ultimately the drive proved too difficult to manufacture).

I had been through enough new model introductions to know that it was dangerous to try to integrate too many new technologies at the same time. If one thing failed, the whole project would fail. But I wanted to prove that Lenovo's ownership of us was not negative.

"Naitoh-san took a risky position," Hortensius recalls. "We had to get a working solid-state drive; we had to get a seven-millimeter DVD. A power envelope didn't exist for that. So we had to figure out a way to cool these components. We had to do a host of things to make this machine special."

We entered the "plan phase" of the project in January 2007 and were due to launch in February 2008.

In October 2007, a problem emerged that threatened to knock us off schedule. Two Asian suppliers of the solid-state drive had not passed Lenovo's quality control tests. We had to drop the suppliers and scramble for alternatives. We faced a crucial Lenovo quality review board in early December, and we could not afford to be delayed. It was a very tough moment. They told us we were not ready to proceed into production.

By now, Hortensius and other top executives had seen prototypes of the new machine, which would be called the X300. They were impressed. It weighed only 3.1 pounds, which was below the target of 3.4 pounds.

We agreed that we did not want to stop development entirely while we lined up a new solid-state drive supplier. Mark Cohen, with whom I had worked on the L40 SX, was now head of operations for Hortensius. He took a risk and ordered "test production" to start in Shenzhen, in southern China, at the same time we were trying to qualify the new supplier of a solid-state drive. We were winking at the quality control board. If we had followed the rules precisely and simply halted all development, we would have lost crucial time. When we did get the new supplier approved, we were able to quickly ramp up production. I don't think the quality control board ever realized that we essentially ignored them.

We planned to start full-fledged manufacturing on January 25, 2008. But on January 15, just ten days before we were to start up, Steve Jobs unveiled the aluminum MacBook Air and declared it the "world's thinnest notebook." In a television commercial, he pulled a MacBook Air out of an interoffice envelope to dramatically make his point. We were worried that this might affect our product launch.

I was not in North Carolina to witness what happened, but a *Business Week* cover story by Steve Hamm reported that after learning about Jobs's announcement around lunchtime that day, Hortensius shouted to his administrative assistant, "Phyllis! Get me one of those interoffice mail envelopes!"

Phyllis Arrington-McGree went through their filing cabinets and found one. She handed it to Hortensius and he slipped an early version of the X300 inside. "It fits! It fits!" he shouted.

We soon realized that we had many more features built into our machine such as a fingerprint reader, a wide-area network card so that the machine could get on the Internet wirelessly, and a CD drive. Hortensius decided to make a YouTube video making fun of Steve Jobs's claims. The video, which can still be seen today, showed that when all the power cords and other devices essential to operating the MacBook Air were included, the combination of devices was so large that it tore apart the envelope. In contrast, the ThinkPad, which had most of the peripherals built in, fit in smoothly.

It was the first time that we went head-to-head against Jobs, the master promoter. It certainly would not be the last.

The X300 was priced from $2,700 to $3,000, which was quite expensive by any standard. It was heavily promoted at the Summer Olympics in Beijing, which Lenovo cosponsored. This was a kind of coming-out party for Lenovo on the world stage. We saw the X300 as a "halo" product that would improve our overall image and reputation. Unfortunately, we introduced it just in time for the financial crisis of 2008–2009. The recession hit and customers retrenched. We ended up suspending production and sales. But we had made our technological statement. And we had set the stage for our fourth and current generation of ThinkPads, the X1 Carbon.

An Earthquake Hits Our New Offices

IBM's facility in the small town of Yamato was designed for three thousand people, and it was filled to capacity at the high-water mark of IBM's presence in Japan. But over the years, as IBM retreated, more and more offices and laboratories were emptied and more divisions were sold off. All my friends in those divisions were gone.

By 2010, we had been with Lenovo for five years but continued to lease space from IBM at Yamato. IBM's commitment to continuing to own the facility, however, was very much in doubt. IBM could have left at any moment. So we decided that we could not wait. If IBM pulled out, we could be homeless people in the street. We started looking for an alternate facility.

Chairman Yang was very supportive of the decision for the employees to move to a better place, with access to transportation, restaurants, and shopping. I wanted to find a building in a location that would not require employees to buy new homes and uproot their families. If everybody had to move their homes and families, that would have cost the company a lot more money.

So the new lab could not be too far in any direction. And we needed room for all our torture chamber devices including very large vibration machines that were a story and a half high. They were very heavy, and we needed a floor that could support them. In addition, some of our torture devices were too noisy to be near an office setting.

As I looked around for space, some building owners were happy to give us office space but would not give us space for our heavy equipment. If we took a warehouse for our heavy equipment, then we would need to create office space inside the warehouse. But conditions inside most warehouses were not very good. They were rather industrial. I didn't want that. I wanted something that communicated a sense of innovation and high quality.

I started looking at buildings in downtown Yokohama, less than an hour away by train for most of our employees. The city was reclaiming many acres of land from the sea and undertaking an ambitious urban development project overlooking Tokyo Bay. The area was called Minato Mirai, meaning "Port of the Future."

It was a very different area from Yamato because it contained large shopping malls, restaurants, a convention center, major hotels, and the Landmark Tower, one of the largest skyscrapers in Japan. There were not many shopping and dining choices back in Yamato, but there were in this area. Plus, this area seemed more connected to the world. You could see planes taking off and landing at Haneda Airport, in the southern part of nearby Tokyo.

I managed to acquire the top two floors, the twentieth and twenty-first floors, of an office building. Our test equipment would be housed in a three-story annex building so that it would not disturb any office workers, whether through vibration or noise.

What would we call this set of laboratories and offices? Many people, both inside and outside of the company, said, "Naitoh-san, please don't change the name. The Yamato name is a brand. No one will ever say that 'Yokohama did this.'" So we called our new facility the Yamato Laboratories at Lenovo Yokohama.

I didn't lose a single person as a result of the move. Soon after we left Yamato, IBM decided to leave. If we had not made our move, it would have been a disaster for us.

Disaster came from a different direction instead. We moved in in January 2011—just in time for the earthquake that made headlines around the world. On March 11, we got hit by the quake at 2:50 p.m., when our offices and labs were at full capacity.

The epicenter hit north of Tokyo in coastal Fukushima and was followed by a tsunami that flooded a nuclear power plant and sent radiation into the sea. I was at my home in Fujisawa and I could feel the quake. It was the biggest earthquake I had ever felt. I was immediately concerned that big coastal cities such as Tokyo and Yokohama might erupt in flames.

Fortunately, Japanese construction techniques have evolved over the years to protect our buildings from earthquakes, which are common here. In the base of our building were parts that moved in response to the earthquake. As I understood it, the building, in effect, floated on a base that could cushion against the earthquake's vibration.

Kumaki, my software ace who had spent years in China, and many others were in the new offices. The electricity went out, but they could still instant message colleagues in the United States on their battery-powered devices, so they were not completely cut off.

Kumaki recalls that the sliding glass doors in every executive's office slid back and forth while the earthquake hit and as the aftershocks followed. That was obviously disconcerting. One of hanging lights on the twenty-first floor swung so far that it hit the wall. That gives you some idea of just how much the building was moving. But the building was designed to move. Other than that light fixture, we did not suffer any physical damage.

The elevators stopped working because there was no electricity. The people who tried to get home had to use the stairs. The streets were jammed with vehicles because traffic lights were out, creating chaos. The only way for some employees to get home was to walk. Many stayed overnight in the building, using flashlights as the source of light and eating instant food we had stored up. I kept checking with my secretary via email and was relieved to find out that nobody had been injured.

People looking out the windows on our twenty-first floor could see oil refineries on fire on the other side of Tokyo Bay. The city of Tokyo was between us and Fukushima, so any continued tragedy could have inflicted enormous human losses.

Our immediate problems were twofold—helping our people and scrambling to fill in the gaps in our supply chain.

When power was restored, we created a command center, or a

war room, to coordinate all our preparations to take care of our people in this hour of need. Even before the quake, everybody had a backpack with water and emergency food, a special thin coat made of material similar to aluminum foil to keep warm if necessary, a telephone charger, a helmet, and a GPS device in case it became difficult to locate their home in the event of widespread destruction. The only thing I thought we needed to add was an extra pair of shoes.

So we checked and rechecked those plans and tried to come up with other contingency plans. If the power goes out again, do we disconnect our computers and servers? If we need to evacuate the building, how do we get everyone out safely? An even greater risk than an earthquake would have been a tsunami of twenty to thirty meters in height. We were at sea level on reclaimed land. The technology that protected our building against earthquakes might not have worked as well against a tsunami.

As we were dealing with our human safety issues, we knew that we also faced a business challenge. Some of the Japanese suppliers of parts for the ThinkPad, which was being assembled in Shenzhen and Shanghai, had gotten knocked out by the disaster. One maker of small electrical connectors, a sub-sub-supplier, was located north of Fukushima and could not continue production.

We also had a major problem with the maker of a certain kind of semiconductor, also located in Japan. Their clean room, which must be kept at certain temperatures and must not have any particles floating in the air, was completely damaged.

We had another war room to track all these supply-chain problems. In many cases, we redesigned pieces of the ThinkPad, found new manufacturers, and got them quickly certified—old friend Ken Yonemochi was now in charge of global product assurance—rather than waiting for a manufacturer to recover.

Even though we are on the cutting edge of technology and many of us live in big cities, we Japanese always remind ourselves—we are profoundly vulnerable to nature.

Chapter 10:
The Yoga and the
Issue of "Consumerization"

We were managing two different processes to make the ThinkPad's next technological and marketing leaps. The first was working with the Chinese labs to develop the Yoga, which I had first seen when I walked through Lenovo's offices in Beijing in 2005. The second was the development and introduction of the X1 and X1 Carbon, the fourth major generation of ThinkPads. Those two rivers of development eventually joined to become one bigger, more powerful river, resulting in the biggest transformation in the history of the ThinkPad.

Some people were critical of us for how long it took to develop really robust cooperation with Lenovo's labs in China. I believe, however, that doing it properly meant it had to be done in a step-by-step manner. Their processes and level of organization had to mature to our level. After all, we had spent years agonizing over how to create all our torture tests and how to accumulate data about each part and each process so that we could make well-documented decisions. The young Chinese engineers were enthusiastic about learning. But it took time.

The collaboration started to get serious in 2011, six years after the acquisition. That year, Peter Hortensius reassigned me as chief development officer for all of Lenovo, not just for the ThinkPad division. This included development of Lenovo's desktop PCs and other machines. I was now going to manage development teams in Japan, China, and the United States. This was our triangle of innovation, and it emerged as an absolute key to our strategy. By now, the Chinese engineers had developed two versions of their Yoga, one with an eleven-inch screen and one with a thirteen-inch screen. As part of my new job, I needed to help them complete these two new products.

I was now on the phone almost every day with key managers on the Chinese mainland. "What are your issues?" I asked them. "You should report your problem this way and in this format." They really listened.

I was particularly pleased with the work style of Audrey Wang, who was very young and who was the project manager for the eleven-inch Yoga. She was not an engineer but she was in charge of ensuring the project ran on time and keeping everyone informed about what was happening. It was a worldwide development job for a worldwide business, and she knew she needed my help because very few people at Lenovo in China had that experience. Just as we at Yamato had learned how to operate in multiple countries with multiple cultures, people like Audrey Wang were now learning how to do that as well. I would wake up and see emails from her sent to me at 2:00 or 3:00 a.m. Ah, I said to myself, the Japanese are not the only ones who work such crazy hours. The Chinese are doing the same thing. They were traveling up the same learning curve. We are not different at all.

The Lenovo teams had several problems. The team working on the eleven-inch Yoga model, in particular, faced big challenges because Microsoft had just introduced a new operating system for mobile devices called RT. It was entirely unfamiliar to us and therefore difficult to manage.

Both teams had difficulty knowing where and how to locate the antennas necessary to provide wireless connections no matter what position the machine was in. You can imagine how tricky that would be if the machine could be used in at least four positions.

An even trickier issue involved the hinges that connected the top half of the computer—Covers A and B—with the bottom half—Covers C and D. The battery was located in the bottom half of the device, close to the motherboard's electronic circuits, and the screen and camera were located in the upper half. They obviously had to be connected. The camera, in particular, required more wires to work in the new Windows system. So more lines had to be run through the 360-degree hinges than usual. If everything was not connected precisely right, obviously the machine would fail. I dispatched my best engineers from the ThinkPad development team in both the United States and Japan to go to China, for weeks or sometimes months, to work on these issues together with their younger Chinese counterparts. We were gradually coalescing and becoming one Lenovo Notebook PC development team.

The eleven-inch went on sale in August 2012, and the Yoga 13 was released in December of that year. They carried the Lenovo IdeaPad brand, not the ThinkPad name. The time was not quite right for the ThinkPad to experiment with an entirely new size and shape.

Think Carbon

While I was working with the Lenovo teams on products for the consumer market, we wanted to maintain the pace of innovation on the ThinkPad itself, which was sold more to businesses and universities. The development of the ThinkPad X1 had centered primarily on driving down the cost from the X300 that had come out in 2008 while at the same time continuing to take advantage of new technologies. The X300 was at the cutting edge of technology, but it had disappointing sales, both because it was too expensive and because it was introduced amid a financial crisis and recession.

The vice president of marketing for the ThinkPad business unit in 2010 was Dilip Bhatia, who became the leading evangelical for the customer. Bhatia spent a lot of his time on the road going to Customer Advisory Council meetings. This was where he gathered key customers and showed them futuristic prototypes of different ThinkPads.

He wanted to learn what they wanted. In a place such as San Francisco, this would involve meeting key people from Cisco, Google, Microsoft, and Oracle. In New York, these meetings would be geared more toward financial and pharmaceutical firms such as Citi and Johnson & Johnson. The message from all of them was overwhelming: make ThinkPads more appealing to a younger generation through features such as better video and sound capabilities.

It was an extremely difficult balancing act. We wanted to maintain the features that customers expected and needed, yet at the same time we wanted to respond to fresh needs and fresh technologies. We ended up adding more backlighting that would allow for better viewing of video, and we added Dolby Home Theater sound for greater enjoyment of music and video. We turned to the American company Corning to provide its Gorilla glass to cover our screens. Corning had come up with this very tough glass to replace plastic screens that had a high breakage rate.

At the same time, we wanted to keep driving down the thickness of the keyboard to allow the machine, as a whole, to be thinner. But we did not want to sacrifice any of the satisfying sense of touch that we had learned so much about over the years. We had started with keys on the L40 SX in 1991 that traveled 3.0 millimeters (about a tenth of an inch) from their starting position to the point that they would "click" and rebound. We could draw a chart showing the ideal force curve of how the key should respond.

As competitive pressures mounted over the years, we started going down step by step from 3.0 millimeters to 2.5 millimeters. Some of our competitors were willing to compromise on the tactile sensations to achieve thinner sets of keys, but we refused to sacrifice any of the key-force curve characteristics. We kept working to improve the suspension mechanism of the keys and we kept experimenting with new materials for the rubber dome that sits underneath the keys.

Gradually, we were able to decrease the keys' travel distance to 2.2 millimeters, then 2.0 millimeters. For the ThinkPad X1, we reduced it to 1.8 millimeters. Partly as a result of this work, the first iteration of the X1 was 17.0 millimeters thick, or about the diameter of a US penny, and it weighed less than three pounds. That thickness was down slightly, but meaningfully, from the X300, which measured 23.4 millimeters in depth.

Bhatia moved up to become general manager of the ThinkPad business unit and started a discussion about how to take advantage of the fact that all our machines since 1992 had used various forms of carbon composites in their bodies. Carbon was lighter and stronger than most metals.

As we debated what iteration would come after the X1, we started experimenting with different surfaces and textures for Cover A. What would make it feel like it was carbon? Should it be slightly rough to the touch? We experimented with making the top layer of carbon fiber visible to customers so that they could instantly understand that these machines were different from other machines that used aluminum. But that solution turned out to be very expensive.

In the final analysis, Bhatia came up with one of those brilliant ideas that are stunningly simple. Why don't we maintain the feel of the machine just the way it is but use the word "carbon" in its name? The best ideas are usually the simplest.

The X1 Carbon was introduced in August 2012 at roughly the same time the Lenovo Yogas were coming out. It was clear that with the X1 Carbon, we were aiming at a sweet spot in the market, a sector that became known as ultrabooks. This was the spot where notebooks, coming from our end of the computing spectrum (the more serious and professional side), competed against various forms of tablets, which had started on the other end of the spectrum (entertainment purposes). Tablets such as Apple's iPads could be used with an optional keyboard, but they could not provide the full productivity that a notebook with a real keyboard could. Most serious users wanted the clamshell style of a notebook. But they wanted the machines to be thinner and lighter.

For the X1 Carbon, we shaved the computer's weight to 3.0 pounds from the original 3.7 pounds of the X1 even as we increased the size of the screen from thirteen inches to fourteen inches. It had eight hours of battery life and the start-up time, a crucial factor for productivity, was cut by as much as half to less than twenty seconds.

We were clearly going toe-to-toe with Apple and everybody knew it. "With the ThinkPad X1 Carbon, Lenovo is throwing its gauntlet down in the already competitive ultrabook arena," Patrick Budmar of Australia's ARN technology review wrote in October 2012.

Others praised it as well. In a review published for CNET, Dan Ackerman wrote, "At first glance, the ThinkPad X1 Carbon looks a lot like other ThinkPads, but in the hand it stands out as very light and portable. The excellent keyboard shows up other ultrabooks, and the rugged build quality is reassuring."

Yet Apple appeared to dominate the media's attention. A month after the release of the X1 Carbon, I conducted an international media briefing in Yokohama. One reporter, Liau Yun Qing, writing for ZDNet, a technology news service, asked me what I thought about Apple. "I'm not ignoring Apple, but I'm not scared of it," I said. That became her headline: "ThinkPad creator 'not scared' of Apple."

Was I supposed to be scared? That seemed to be her implication— that I was crazy for not being scared. My point is that many in the media have consistently exaggerated Apple's capabilities even when we had a clear technological lead.

The Rivers Join

By 2013, the stage was set to allow the two rivers of innovation to start flowing together, one being the Yoga and the other being our sacred ThinkPad brand. We were going to offer the most radical alternative to the traditional ThinkPad form factor since it had been introduced.

We had a ThinkPad Convertible model, but it only operated in two modes, the normal clamshell mode and the tablet mode. After we had worked on the Yoga for a period of time, we realized that its four-way capability was better than our convertible.

But before we could incorporate Yoga elements into the ThinkPad line, we needed to fix another problem. When the machine was placed in a tablet mode—meaning the top half of the ThinkPad Yoga was folded backward and met Cover D on the bottom of the computer—the keyboard would be facing downward, toward a table or desk or some other hard surface. Because the keys rose above the level of the frame, they were exposed. The keys could be damaged by any number of random movements.

Mechanical design leader Agata started to develop answers with the help of Mr. Owl Wing Nakamura. The team first examined what it would take to apply enough force to cause the keys to recede into the frame of the keyboard, thereby protecting them.

Each key cap sat on two crossed mechanical arms and a rubber spring. Team members calculated that it would take thirteen pounds of pressure to cause one key to recede into the frame. Because there were one hundred keys, arranged over six rows, a huge force would be required to press down all the keys at the same time. How could we generate such a force? And even if we could figure that out, we would have to have a bigger, heavier computer to support the force. We didn't want to go down that path.

The team came up with a new idea. Why not do precisely the opposite and make the frame rise to the same level as the keys? The keyboard would then become perfectly flat and therefore the keys could not be damaged. In something that would resemble a bizarre Rube Goldberg design, the mechanical guys figured out how to etch small grooves in the two hinges and then connect them to two slide-like bars that would run inside the bottom half of the computer along the two edges, left and right. These sliding bars were, in turn, connected to steel wires that ran from left to right, and right to left, through the frame of the bottom half of the computer. The way they described it to me, when the machine was opened all the way into the tablet position, the steel wires would automatically raise the frame to the desired level. No motor was required.

When I first heard about this, I said, "Come on. Are you guys serious?" I just couldn't believe it would work.

But they kept developing the idea in very detailed ways. It required incredibly precise manufacturing practices, with a system of strict controls. All the tolerances had to be just right. In the final analysis, it worked. We called it the "lift and lock" system, and it was an industry first. We were taking innovation to a whole new level.

Thanks to these efforts, we released a ThinkPad Yoga with a twelve-inch screen in Europe in September 2013, and in the United States in November of that year.

In 2014, we updated that offering with the ThinkPad Yoga fourteen-inch screen. And we hit hard in 2015 with four more versions of the ThinkPad Yoga. It was a real product blitz.

But our boldest statement was the fourteen-inch ThinkPad X1 Yoga, which was loaded with everything we had. It was unveiled in January 2016 at the Consumer Electronics Show in Las Vegas. It could be operated in four positions and weighed only 2.8 pounds.

On the ThinkPad X1 Yoga, we started offering the PC industry's first organic light-emitting diodes (OLED). This technology was pioneered by Samsung Electronics and other big South Korean electronics companies. They had been using the technology in smartphones and in tablets, but we adapted it to ultrabooks.

The word "organic" in the name of the technology does not imply that there is any living tissue in the screen but rather that carbon, an organic material, is used. When electricity is applied, each pixel emits light and color. Unlike LCD pixels, OLED pixels do not require a backlight and the screen can therefore be much thinner.

What most people notice about OLED is the brilliance of the color and the stunning clarity of the images on the screen, which looks the same when viewed from any angle, unlike some of the big television screens.

The problem with OLED screens, which we had to address, is that icons displayed on them will start to burn into the OLED material if they are left on the screen for too long. This problem was called "image sticking." Each pixel of the OLED screens could be degraded if excess energy were injected into it for too long a period of time.

We had to come up with both hardware and software countermeasures. In hardware, we learned how to measure the cumulative stress that each pixel was experiencing. The software piece of the solution was offering users the option of a "dimming task bar" and a "dimming background." These features basically maintained full brightness only in the window that a user was watching or working in. They minimized the brightness of other background windows and the task bar at the bottom of the screen. When a user switched to other windows or returned to the task bar, full brightness of the screen would shift with the user. This shifting of power helped the OLED screens minimize energy consumption and reduced the problem of image sticking. We had patented this software more than ten years earlier for the purpose of saving power on LCD screens, but we realized it was far more valuable when used with the new OLED screens.

The X1 Yoga also incorporated a pen called an active stylus that can be used to draw sketches or concepts on the screen, or to make corrections to a document. Those images can be captured and transmitted to others, which could create a whole new avenue of collaboration among users. It is like using a physical whiteboard to sketch out new ideas except that the ideas can be shared electronically with others who may be far away. We had introduced one product with a pen many years ago, and it was not a success because the technology was not ready. But this time, I predict the pen will find broad acceptance. We fixed other problems as well. We created a space inside the computer where you can slip in the pen for simultaneous storage and recharging. Where to store the pen and how to recharge it had long been pain points for customers trying to sketch or draw on their devices, but we now had eliminated those problems, opening up an exciting new way for people to use computers and to communicate complex ideas.

Consumerization and the Design Challenge

As I have described, ThinkPads were traditionally sold through universities and corporate IT departments, and those buyers valued security and reliability. They were interested in the total cost of ownership of the devices and wanted them to be as uniform and as security-conscious as possible. Many technology-minded individual users also bought our products and they were very enthusiastic supporters, but they were not the majority of our customers, at least not in the United States.

With the rise in popularity of many devices from Apple, however, people inside these enterprises and institutions about five or six years ago started demanding the right to make their own choices about which devices they used on the job, at home, and on the road. The balance of power inside large corporations and schools seemed to shift—individuals did not want to use devices that a senior vice president in charge of IT provided to them. They wanted to use their personal devices. This seemed to represent a direct challenge to us.

We had always considered our designs very rational while it seemed to us that Apple had put a greater emphasis on the emotional connection. One reason Apple's MacBook had such a powerful emotional connection with users was the shape of its aluminum body. Apple used a technique called computerized numerical control (CNC) to shape each body from a block of aluminum. That helped create a unique shape with smooth curves, which appealed to many people.

But for people such as myself who grew up in the IBM manufacturing mind-set, this very expensive method was beyond our wildest dreams. We were much more focused on driving down costs. I have always been truly impressed with what Apple did with its design and manufacturing processes. They understood emotionality. We understood rationality.

But now the lines were blurring. We had long debates internally about whether we should "consumerize" the design of the ThinkPad in response to Apple.

As it turned out, the real challenge from Apple products inside the enterprise and inside the university came from the iPhone, not the MacBook or the iPad. People who loved their iPhones outside the office also wanted to make business calls on them and send quick emails. People who used Apple's notebooks, ultrabooks, and tablets such as the iPad Pro were less eager to use them for business purposes because they were not well suited for handling serious documents and spreadsheets.

Still, we worried about the threat or potential threat of people leaving behind their ThinkPads for a more aesthetically appealing device. We debated how to inject more emotion into the ThinkPad design. With the X1, we started tapering the edges of the machine, moving away just slightly from the chunky, squared off contours that the ThinkPad had long displayed. That adjustment was well received. But we tried to inject more pizzazz into the second iteration of the X1 Carbon and that was not as well received.

The machine was still black, but we added a little silver powder to make it a more interesting design. Over the course of the next year, we learned that we had gone a little too far. I heard a lot of customer feedback, especially from the members of a China ThinkPad fan club, whom I met. In this face-to-face meeting, they told me that the greatest black in China is the emperor's black, which means pure black. They were very insistent that they did not want any other shades of black. They wanted us to return the ThinkPad to its all-black color. We did, in fact, decide to go back to the emperor's black. Who could have anticipated such sensitivity?

We will keep working on ways to inject more emotion, but we have to do it in our own way. I cannot tell you exactly what we are planning because it would give away secrets to our competitors. But you can expect some new visual design elements. And they will be done in such a way that does not anger our hard-core supporters around the world.

There is a deep emotional connection between people and the design of all the objects we use in life, but it seems the connection is particularly strong when it involves our communications and work devices. All this is related to the trust our customers have in us. The word "trust" is one that I emphasize all the time to everyone working with me. The other concepts I keep hammering away at are that everything we do has to help achieve "customer success" and enhance the customer's "productivity." These are more than just empty corporate slogans. These are the foundations of our work philosophy. For everything we do, we need to be able to explain to our customers why it is good for them. Design is a deep part of that engagement. It takes years to build trust, but the trust can be broken overnight if we stray too much from our core mission.

An Army of Innovators

As a result of our engagement with Lenovo's labs on the mainland, over the period of a decade, we now offer completely different form factors for the ThinkPad and we have incorporated the very latest technologies such as OLED and the pen. Lenovo as a whole, including the ThinkPad, offers the most advanced selection of notebooks and ultrabooks on the market, and that is the one sector of the overall PC industry that is still enjoying strong sales. Our innovation triangle has worked and set the stage for further gains.

The reason I expect further gains is that our innovation model is not dependent on one individual. We have research teams looking out over the five-year horizon for emerging technologies. Then we have development, which is considered a different function. These people are in charge of looking out three years to verify how those new technologies that have been previously identified are maturing. Each development organization, in turn, has an innovation team that takes an idea that has been developed and tries to incorporate it into an actual product. It is like a relay race in which we pass on the batons.

So a whole chain of people is involved in our innovation model of multiple cultures and geographies and ages and worldviews. Our model will not die when I pass from the scene. I have built an army of innovators who will keep marching forward long after I am gone. In fact, I predict they will create a bigger wave of innovation than I was ever able to achieve myself.

Chapter 11:
The ThinkPad's Impact
on the World

As you have seen, the ThinkPad helped change the way we understand our planet, and the stars and planets around it. It encouraged individuals to explore in new ways. In the broadest sense, the ThinkPad helped change not only the way people collect information but also the way they store and access that knowledge. It used to be that knowledge was stored in libraries or central repositories, and the seeker of truth had to make a pilgrimage to those centralized institutions. Yes, millions of people used physical encyclopedias, dictionaries, and thick Yellow Pages containing business details and phone numbers. But those reference tomes could not possibly contain all human knowledge, as that knowledge began exploding.

The idea that information could be accessed remotely from any point in the world is a radical concept when viewed over the course of history. The Internet enabled that, but the ThinkPad, beginning in the early 2000s, helped facilitate the most robust exchange of documents, reports, PDFs, technical designs, spreadsheets, and other serious works of knowledge over the Internet. There were other devices, to be sure. But the fact that more than one hundred million ThinkPads were sold and put into use means it was an essential player. Its role in developing Bluetooth technology, now embraced by tens of thousands of companies, also was key. Taken as a whole, these devices and the Internet—this ecosystem—changed the very architecture of human knowledge.

The ThinkPad also helped change the way humans make decisions. Pause to consider. It used to be that most major decisions were made by relatively small groups of people who met face to face. They were at the top of decision-making hierarchies. But now because of the Internet ecosystem, of which the ThinkPad is a major element, more people who are not physically present can be involved in decision making. The era of the network has been born. In many cases, those in power have clung to their hierarchies to insulate themselves from criticism or challenge from their subordinates, or from constituencies outside their organizations. But the technology offers a clear alternative.

This power of the network has changed the way work is done and where it is done. It helped change the way doctors and nurses and medical clinics, for example, store patient data and seek to improve health care. The data can be stored in a central location, and each doctor's clinical observations and prescriptions can be seen by other doctors and nurses within the practice on their ThinkPads. That cuts out a tremendous amount of duplicated paperwork. The ThinkPad changed the way police departments manage themselves because many deploy ThinkPads in patrol cars, to the chagrin of any speeding driver. A police officer can access a driver's entire driving history from the state's department of motorized vehicles and also check for any outstanding warrants in just a few seconds. Highly mechanized farmers have ThinkPads in the cabs of their planting and harvesting equipment to help them identify which sections of their fields should receive the most fertilizer, for example, as analyzed on the basis of satellite images. Crews at a construction site can keep better track of where all their materials are and when delivery can be expected by using their ThinkPads to link up to the systems of their suppliers.

In effect, there has been a democratization of knowledge. We know this ecosystem can be abused by people who want to promote false news stories, or it can be abused by people seeking to hack into other people's systems and steal data. We know there are downsides, but the upside has been much bigger—our relationship to knowledge and decision making has changed forever. It is up to us to make the best use of this new power.

Perhaps the deepest sectoral impacts were in the worlds of education and business. Those two sectors bear examination in greater depth.

Changing the Way Kids Learn

Colleges and universities were the first educational institutions to embrace the ThinkPad. Harvard Business School was the first to buy the predecessor L40 SX in 1991 and then the first to buy the ThinkPad in 1992. But it wasn't until the early 2000s that most universities got on the ThinkPad bandwagon. Over these years, students used the machines much as one would expect. They conducted online research for papers, collaborated with other students on joint projects, and engaged in email exchanges with professors over grades and such. The class registration process, which used to require standing in physical lines for hours on end, was greatly streamlined as it moved online.

The ThinkPad also gradually helped change the way professors taught their classes in fields where extensive memorization was required. It no longer made sense, for example, for medical schools to demand that students memorize anatomical details, or for chemistry programs to force students to memorize the makeup of molecules if that information were readily available online via a ThinkPad or a similar device.

But college students are already at a certain point of maturity and they may not have absorbed the full impact of the new age of mobile computing. The ThinkPad "basically allowed you to write very different kinds of cases," Harvard's McFarlan recalls. "You could have exhibits available in electronic form. You could build video into it.

Students could gather in small groups and build PowerPoint presentations. Suddenly there was less paper. People were sitting around with their ThinkPads sharing information and comparing notes."

That give-and-take was always built in to the Harvard case study methodology so although the ThinkPad was helpful, it may not have been transformative. "The quality of thought remained the same but the vehicle through which it was presented was absolutely changed," McFarlan explains.

The ThinkPad may have had a greater effect on children at earlier, more formative stages of their lives. It may be that the ThinkPad ultimately had greater impact at the high school level than at the collegiate because of the way it deepened interactions between teachers and students. In some cases, it changed the nature of education.

Consider the case of Cardinal Gibbons High School in Raleigh, North Carolina. Lesley R. Coe, a former Spanish teacher from Long Island, New York, was hired to become the high school's director of technology in the fall of 2012. It was, and remains, the largest Roman Catholic high school in the region with 1,485 children in grades nine through twelve. The school already used ThinkPads for teachers and had other desktop machines for student use.

But the school's goal was for everyone, students and teachers alike, to be equipped with a personal device. It was Coe's job to figure out which device. "If anyone had asked me at the start of planning, 'Hey, are you guys going to do a laptop?' I would have said, 'No,'" she says. "I have spent a lot of time in schools seeing all the challenges with laptops. The laptop is the hardest computer to manage and support technically."

Coe went through an elaborate research process to assess the needs of students. The school aims to prepare its students for college, so equipping them with computer skills was seen as essential to their development.

Coe studied a variety of notebooks. Initially, she and her superiors rejected the ThinkPad Yoga. "There were some flaws in the physical design that were not going to be great for high school kids," she recalls. "When you flipped it over, the keyboard was still exposed. That was not going to be good for us. Keys were going to come off. We would be replacing keys all day long. It wasn't going to meet our needs.

Time passed and the deliberations continued. It just so happened that one of the parents at the school was Luis Hernandez, who was the latest vice president and general manager of the ThinkPad business unit, also located in Raleigh. "Luis walked in one day with a ThinkPad Yoga," Coe recalls, "and we said we don't want it because of the keyboard."

Hernandez produced a new version of the machine. "Will you just watch?" he asked. Hernandez flipped the top of the machine down and around to create a tablet, and the level of the frame rose to protect the keys, as we engineers had worked so hard to achieve.

That won the day. "It could be used in tent mode and students could sit around and watch it," Coe says. "In tablet mode, I saw how the kids had sheet music up on the screen and were using it to play the piano we have here that anybody can use. I was like, 'Wow, I had never thought of that.'" The touch screen features also were advanced and Coe knew students were very oriented toward the use of touch in engaging with computers.

After extensive research and professional development work with the teachers, the school equipped freshmen and sophomores with ThinkPad Yogas beginning in the fall of 2014. In each of the next two years, it equipped incoming classes of freshmen and will do so again in the coming school year. At the end of that three-year rollout, all students had slightly different iterations of the ThinkPad Yoga.

One challenge in using any computer in the classroom was the potential for a student to become distracted during class. Teachers quickly learned that when they want the attention of students, they should specify, "Okay, everybody, lids at forty-five." That meant the top half of the ThinkPad Yogas should be placed at a forty-five-degree angle, making it impossible to actually use the machine but leaving it powered on. Then when the teachers wanted students to use their machines, they could specify it was time for ninety-degree angles.

Those issues aside, the machines seem to have a deep impact on the quality of the educational experience. On a recent snow day, for example, students in Advanced Placement (AP) courses who approached test deadlines set by the College Board were able to have their classes conducted in Google Hangouts, thanks to their connected ThinkPad Yogas. That way, they remained on pace with the new material they were expected to master. They didn't have to actually be in a classroom.

But at a deeper level, the ThinkPad Yoga seems to have intensified contact between teacher and pupil, particularly in English and other classes such as social studies in which writing is an integral part of the work. The school's English teachers have become the biggest fans of the machines.

A teacher can provide a link during class that all the students can access at the same time to retrieve a particular document. The teacher is able to see who reads and comprehends best, judging on the basis of what they write and present as their conclusions. Because of the collaborative features of Google for Education, a single student or a group of students can be working on a paper, and a teacher can see what they are doing remotely. He or she can provide feedback in real time as the work is being done. The teacher can leave an audio link expressing an opinion or can highlight a passage in the paper that needs particular attention. The teacher also can have an online chat in real time with the student or multiple students. Different teachers prefer different styles of engagement.

"Everyone makes the assumption that technology is making it easier for teachers but it is actually making it harder because it is more intensive," Coe says. "The quality of their interactions has increased. If you are grading a stack of ninety papers and you're hand-writing your comments, you are limited in terms of time and energy. Digitally, you can do it any way you choose. There is a record of the interaction that went into that paper. The teacher is giving more meaningful feedback."

Delivering that meaningful feedback in a timely way "is always the challenge with education," she adds. "If the workload is so great that the feedback is delayed, the connection is somewhat lost." Immediacy is crucial if someone is to truly learn.

The ThinkPad Yoga's pen also has the potential to change the way math and other subjects are taught. The first year, there were problems with the pens, such as where they would be stored when the computer was not in use. Pens that were not stored properly could be lost or damaged. So the school that year declined to buy pens for the computers. But the second year, the machines had places to store the pens. Hernandez once again demonstrated the improvement and persuaded the school to buy into the concept.

That pen was "fantastic," Coe says. "I had a teacher who writes math problems all day. She was my guru. We had purchased every stylus that Amazon.com sold. We tried everything. We had some okay solutions. But one day she came in to my office and I said, 'Try this pen.' She was sold. It responded the same way as if you are writing on a piece of paper."

"One last positive contribution is that kids are learning how to manage computers," Coe says. This is not part of any formal curriculum, but students are learning how to solve problems and troubleshoot their computers. "I can't think of many jobs outside of here when they leave that are not going to require some knowledge of computers like this," she says. "They need to care for their machines. They have to reboot if they need to install updates. If the computer is not responding, they need to reboot."

It is not only the students' technical skills that improve. Learning how to manage their computers is also "helping kids with their social skills, with the skills they will need to enter the world. They are learning if they have a problem with their computer, there are things they need to do before they come for help. We are preparing them. They are going to be much more future-ready."

And "future ready" is what education is all about.

"What Are You Doing Here?"

It's impossible to measure the impact that the ThinkPad had on the world of business because so many technological and business trends occurred at the same time. But there's no doubt that the machine helped thousands of companies achieve major gains in productivity and sales. That translates into many billions of dollars of wealth for employees, shareholders, and other constituencies. In terms of dollars and cents, or yen or euros, the business sector is where the ThinkPad had the greatest cumulative effect over the years.

The first major brand-name company to use the ThinkPad was Coca-Cola, which needed to manage a large footprint of bottlers, distributors, and retail outlets around the world. It still is a major customer. Other early adopters were consulting and accounting companies, which depended on sending their experts out to spend time with customers. Pharmaceutical companies also were early adopters because they had large sales forces that met with doctors and hospitals in hopes of persuading them to buy their drug products. They could now show what a product looked like and offer extensive technical information. One company, Merck, asked us to alter the color palette on its ThinkPads so that one of its pills would be more visually appealing. These sales representatives were the first road warriors.

Michael B. Spring, a professor of information science and technology at the University of Pittsburgh and a self-avowed geek who bought his first Compaq luggable in 1982, is an expert on how technology changed the world of business. He is also a more neutral observer than I can be. Spring believes there were three major phases of how companies embraced information technology, including the ThinkPad.

His theory starts with the observation that publishing was traditionally any Fortune 500 company's second-largest business, coming right after their core business. The reason was that an average of 6 percent of their gross revenues was spent on publishing, whether it was internal memos and reports, product manuals, expense reports, press releases, or annual reports. There was traditionally an avalanche of paper inside large corporations.

Beginning in the 1980s, these companies started using information technology, but their productivity actually declined during that decade. They were spending money on new IT systems but had not yet begun to discard the old paper-based systems. They were spending 6 percent of their revenues on paper and 4 percent on digital systems, for a total of 10 percent. "While you are making the transition from paper to digital, you have 10 percent of your spending that is not productive," Spring argues. People still kept their old-fashioned filing cabinets even though they were starting to save data on their file servers because they were not confident the servers would not fail. They ran two systems in parallel.

In the 1990s up until 2000, the ThinkPad and other notebooks could be taken into the field. That was a breakthrough because sales representatives and consultants could carry information with them and present it visually. But the notebooks were not connected in real time to the head office because they were not wireless. They still needed modems to connect and those connections were still slow. So a sales representative could tell a doctor what the price was of a certain drug and how that drug performed, but it might not have been the very latest information. It might have been accurate when the sales rep closed his notebook, left the office, and then traveled to meet the customer. But if data had changed in the meantime, the sales rep's information on his notebook was out of date. Overall productivity gains were still very limited.

It was not until after the year 2000 that the full potential of mobile computing in the business world started to be realized, Spring says. He argues that three major trends hit at the same time—the cost of storing data plummeted, as did the cost of processing the data. And network bandwidth exploded. It was something like an arms race. As Spring describes it, "Every year, someone would say, 'We have ten times more bandwidth than we can actually use. But then came a semiconductor that could process data ten times faster and we would find something to put on that network that took up ten times more capacity.'" And so it went.

At the same time, the training of secretaries and sales reps and accountants and everyone else began to catch up with the sophistication of the machines. People started figuring out how to maximize the capabilities of their ThinkPads and other devices.

Add it all up, and companies began achieving major gains in productivity. Old paper-based systems were eliminated. Remote workers could carry their ThinkPads and other devices and be connected in real time to headquarters. "The ThinkPad set the stage for having that ubiquitous computer architecture, where an accountant or other business person could have access to data in real time," Spring says. "The ThinkPad computer was the first big tsunami in that direction."

One person who lived through this upheaval, on the inside, was Jim Steele, the young aide to IBM CEO John Akers who helped deliver the first ThinkPad to President George H. W. Bush. He spent twenty-three years with IBM in postings to New York, San Francisco, and Tokyo.

After a brief stint at Ariba, the Internet commerce company, he landed at Salesforce.com in San Francisco in 2002. Salesforce.com, which provides Internet-based tools to help companies better manage their customer relationships, is run by Marc Benioff, who is known throughout the technology and corporate worlds for his swashbuck-ling and sometimes provocative style.

When Steele started there, he asked for a ThinkPad from the company because that was what he was used to. He had virtually grown up using a ThinkPad, and it was the notebook he was the most comfortable with. The company issued one to him as requested be-cause he was going to be on the road a great deal meeting customers. He was head of sales.

One day in his first week in the job, Benioff came to Steele's of-fice and asked, "What are you doing here?"

"Well," said a slightly flustered Steele, "you hired me." Steele thought Benioff was kidding around.

"Yeah, I know I hired you, so what are you doing here?" Benioff insisted. "I don't see any customers here."

Steele was an accomplished sales executive and had experienced the concept called "hoteling" in the 1990s when sales people first started realizing they could spend more time on the road. But he ad-mits now, in retrospect, that as of the early 2000s, there were even fewer reasons to be at the home office. With Benioff's prompting, he realized, "I could be out in the world taking my office with me."

Steele told Benioff that the CEO would not be seeing him very much. He took his ThinkPad and Blackberry phone and started es-sentially living on the road all over the world. He logged so many miles that he became a top-tier traveler for United Airlines.

"What happened in the early 2000s was that I realized my office could be virtual," Steele explains. "I could work from a hotel room, from an airport. It changed the way I did business. My face time with customers doubled. I didn't have to find an excuse to be back in the office."

In the first couple of years, he spent 200 nights a year in hotel rooms somewhere in the world. Eventually it reached 230 nights a year. His Twitter handle became Roadwarrior247, with the 247 referring to twenty-four hours, seven days a week. "I made thousands of customer calls face-to-face and recorded them on a mobile app that Salesforce had," Steele says. "All my customer interactions and visits, I would record through my ThinkPad." Those recordings could be analyzed back at headquarters for any other clues or insights about customers.

Steele reported back to Benioff every month and to the entire management team every quarter. He reported on all his customers, what they liked about the Salesforce product and what they didn't like. He reported back on competitive threats that might be emerging from other companies and other countries. He described new opportunities and also pointed out situations where Salesforce had dropped the ball and missed an opportunity.

"I wrote all that up for years on my ThinkPad. That's how I communicated. I never felt I had to attach myself to a fixed office."

Did he create wealth? He believes he did. The company's sales grew from $20 million a year when he started to $5 billion in the twelve and a half years that he worked for Salesforce. That supported many jobs and created gains for many shareholders. "My road warrior status had a lot to do with it," Steele argues. "I was out there as the customer-facing head of sales. I had to be out there. I had to show the customers how I used Salesforce myself. I would do that on my Blackberry but also a lot on my ThinkPad. Over time, I gave hundreds and hundreds of demonstrations to customers."

How much wealth did the ThinkPad create if thousands of companies were using it in similar ways? "It has to be trillions of dollars," Steele says. "You think about what it did—it untethered you from the office. For anyone who's face-to-face, calling on customers, this is their arsenal, their war chest. If you don't have a laptop and a smartphone these days, you can't survive as a sales person. You have to have the right information at your fingertips. That's what made it so valuable. You wouldn't have to say, 'I'll go back to the office and research the information.' The real-time value and responsiveness dramatically improved the productivity of sales people. The ThinkPad was the original. That's where it all started."

Of course, the ThinkPad was used not only by sales forces. It is being used as a key part of the research and development activities in which scientists enter their research data in computer systems that are intended to make their work visible to other teams of researchers. At Bristol-Myers Squibb research facilities in New Brunswick, New Jersey, for example, scientists enter data from all their research projects into their ThinkPads, which links them to a centralized database of knowledge about what other Bristol-Myers scientists have done. It is called an Electronic Lab Notebook system. The goal is to prevent one scientist in one arm of the sprawling company from repeating the mistakes that another scientist has already made or to prevent her from spending time and money on an experiment that another scientist has already proved to be viable. The vast majority of pharmaceutical companies use these systems to cut out waste in the research process and hopefully speed up the development of new drugs.

If you add up the specific impacts across industries, it's no exaggeration to argue that the ThinkPad changed the face of business. In fact, the ThinkPad and the greater mobile computing revolution of which it is a part have had a deep impact on almost every sector of human activity. And it's not over.

Chapter 12:
The Future of
Mobile Computing

I am near the end of a very long career in which I have witnessed unbelievably rapid developments in technologies, perhaps the most rapid in human history. People today take it for granted that they can use a phone or a notebook that has a bright color screen and connects them wirelessly to the whole world and to all their personal content and contacts. Video flies invisibly through the air. Each step along the way of creating this reality—driving down the size of batteries, improving screen resolution, increasing the speed of computing, building in antennas to connect to wireless networks—has been a battle fought over a period of many years. All those struggles set the stage for what Google, Facebook, and Amazon offer so seamlessly today.

My current and final job is scouring the world for the latest technologies that might be incorporated into our computers. That allows me to see over the horizon just a bit. As a result, I know that the pace of technological change will continue, and perhaps even accelerate, long after I have retired. The next generation of innovators has grown up with the Internet whereas for older people, it arrived at some midpoint in our lives. We can expect that younger people, having absorbed the Internet into the very fabric of their existence, will come up with ideas we can scarcely imagine.

I think I can talk with some confidence about things that will happen within five years. I will let my younger colleagues share their ideas about what can happen in twenty years.

Devices

I think that for the foreseeable future, the majority of professional people will carry two devices. I don't see a convergence onto one device happening very soon—if ever. The reason is that most people need one device with a full keyboard so that they can write and manage documents or work on spreadsheets. Then people need a phone or phone-equivalent for making calls, taking and exchanging pictures, listening to music, sending quick emails, texting, chatting, and the like. Those devices are overwhelmingly oriented toward consuming information, not generating it. Serious professionals will find those devices lacking.

One of the things I've learned—the hard way—over the years is that you cannot introduce a technology before it matures or before people are ready for it. I mentioned earlier that, in 2001, we introduced a pen that could write on a pad of paper. A digitizer then captured the strokes and transferred the writing into the computer's memory. The system was called TransNote. But it failed in the market. The user experience was not great. And people weren't ready for it.

Similarly, I don't think the market is ready for watches or glasses as their primary points of contact with the rest of the world. They may be considered cool and early adopters will rush to them, but they are not yet delivering enough value to the user. Take the concept of a computerized watch. This has been a dream for engineers for many years—a nice display with computing power, storage, and a battery in a small, neat package. There are several e-watches on the market, and I would love to buy one.

The crucial hurdle that any watch-as-computer has to leap over is as follows: Why would you give up your favorite watch, perhaps handed down in your family and therefore of some sentimental value, if all the information you need and all the communications capability you need is on another device in your pocket or pocketbook, just inches away? The e-watch would have to offer a quantum leap in value and utility before it would make sense for millions of customers to buy it and put their beloved traditional watches in a drawer.

The problem with existing e-glasses is the size of the screen. You can see only partial images, and you cannot type on it. That means it does not communicate well, at least not yet.

Regarding notebook and ultrabook computers, I know that they will continue to get thinner and lighter, but the keyboards and displays will still need to be a reasonable physical size. My slogan for the next five years is "Carry Small, Use Big."

Several new technologies are coming. One just beginning to emerge is the use of projectors to display a document or a file from a computer on a blank wall or blank screen. These projectors are going to be embedded on the backside of notebooks and ultrabooks, in the territory between Covers C and D. That means it may be possible to create machines that don't have screens at all. We may not need Covers A and B.

Before that can happen, however, we need to find a way to make the projected images brighter and denser in resolution and for the projectors to use less power.

In this five-year time frame, it may even become possible to create a very flat, flexible piece of plastic. You might even be able to roll it up and put it in your pocket. You could take it out and unroll it—and there would be a keyboard. We are trying to figure out a way for such a keyboard to give the user the feeling that he or she is using a real keyboard. The tactile sensation of typing is so important. We might be able to equip this piece of plastic in a way that it could send tiny electronic signals to the fingers to recreate the feeling of a real ThinkPad keyboard.

The quality of screens, meanwhile, will continue to improve. We introduced organic light-emitting diodes on ThinkPad X1 Yoga, as I mentioned earlier, and many people were amazed by the clarity. OLED screens are thinner than the traditional LCD screens, which means we will be able to continue driving down the thickness and weight of the machines overall. OLED screens also can be curved, which opens up interesting possibilities such as entirely new shapes for computers.

Connectivity and the Cloud

There was a time not long ago, believe it or not, when I was not sure that we would be able to enjoy almost universal connectivity. In the fifteen or so years that wireless technology has been evolving, it seemed to me at one point that it would be limited to certain areas in certain nations for certain types of users. You would never be connected in tunnels, for example, or in remote rural areas.

But I have been proven wrong, and that has big implications. It means we can assume near-universal connectivity. If your device is always connected, that means cloud computing can provide more computing power to you and safeguard more of your data. Cloud computing, for anyone who has lived on another planet in recent years, is the ability to wirelessly connect with networks that, in turn, connect to giant server centers run by the likes of Amazon Web Services, Microsoft, Google, and IBM. These cloud functions are appearing on many personal devices already. You can back up photos, music, video, or documents in the cloud.

Right now, some people think of the cloud as just that—a backup. But what if the service became so good and our confidence grew so high that we could leave all our content in the cloud? It could be much the same for intelligence, meaning software. If we can connect wirelessly to programs and apps that are running in the cloud, why do we also need to have those capabilities residing on our devices?

The implication is that devices will need to have great screens, great keyboards, and great ease of use, but they won't have to store as much data or as much software. That opens up great new possibilities. One is that you can access your information from different devices. Once you move it to the cloud, it is synchronized and is identical no matter what device you choose to access it with. If you change an address on one device, it will automatically change when viewed from another device. If you change bookmarks on one computer, it will appear the same way on another device.

Another implication is that the Internet of Things can become a reality. Every device—whether air conditioners or household appliances or security cameras—can have intelligence and be connected. Voice communications with these devices should work better than with an actual computer because the range of instructions is much smaller. You could wake up in the morning and say, "Curtains, open up."

Caveats obviously apply: The security of the device, the network, and the cloud are all incredibly important. You need to trust the cloud before you use it. The technology is already there; it's just a question of building human confidence in it. The other issue is that many Internet-of-Things devices are appearing but each needs to be connected to the supplier's own cloud network. How do you know which networks you can trust and which ones you can't? How can you remember all the passwords? The goal should be to create a single password that connects you to everything. Once you have a single Lenovo ID or a Microsoft ID, and your identity is established, you should be able to move from device to device seamlessly.

The Interface

Voice recognition is getting much better, partly because of the much larger computing capacity offered by the cloud. I see significant improvements with Apple's Siri, Microsoft's Cortana, and Amazon's Alexa. I think they will become more important means of using a computer.

There are many challenges in using voice as the primary interface because of the sheer complexity of language. But I am optimistic.

Touch continues to be a great user interface because it can be so intuitive. I predict many children today will grow up using only touch for many functions. It's probably a generational thing—older people will continue to use a keyboard to engage with their productivity device but over time, touch will emerge as a primary interface for younger people.

One means of engaging with your device that we are experimenting with is the gesture. Your body is a big capacitor. When you touch a point on the screen, the capacitance changes, meaning the machine detects the touch. But if we increase the sensitivity of the screen, which is not hard to do, it can also detect your hand from several inches away. In effect, near-range sensing could be achieved by extending touch sensors.

The problem with this is that if you gesture in a slightly different way, the machine could be confused. Or it might interpret random hand movements as gestures that require action on its part. We are still working on this technology. The question is not whether it can work or not; the issue is whether we can create a good user experience. One possible solution would be to use a camera built into the notebook or other device to capture images of your fingers in different positions. That would allow the machine to capture much more accurate information about what gestures the human is making. This method will require more processing power in the computer, but that problem should not be insurmountable.

A related means of engaging with a computer screen is the gaze. Within two or three years, the computer will be able to recognize where your eyes are. You would have to concentrate on an icon, say, for a period of time before the computer concludes that you want to click on that icon. It would then open the application automatically for you. If your eyes are clearly scanning and are not resting in a single place, then the computer would not react. There are still issues with this means of input, but they can be conquered. This technology would be particularly helpful to people who are paralyzed.

Sensors and Cameras

There is an absolute explosion coming in these fields. Sensors are getting smarter, faster, and smaller, and there are more kinds of them. Most smartphones already have a compass and can sense North, East, South, and West. They know the angle if you lift the phone over your head or when you lie down. And of course, they offer location services through the GPS capability. Those capabilities are coming to notebooks and ultrabooks as well.

In addition, sensors in upcoming notebooks will be able to detect and react to other networks nearby. This is an important security feature. Your notebook will be able to tell whether you are in a home environment or in a train station or airport. It might shut down in a public space if it senses that unknown programs or unknown networks are seeking to communicate with it. The computer will also know if it is moving, which will provide it with clues as to the nature of the networks it encounters.

Cameras also are poised for a great leap because they soon will be able to capture three dimensional (3-D) images, not just two dimensional. To take a complete 3-D picture of an object or a person, you will walk around them in a circle and the camera in your device will capture every angle and produce a 3-D image.

Artificial Intelligence

Artificial intelligence (AI) in the form of IBM's Watson or Deep Blue—or in the form of Google's DeepMind—is obviously powerful. But what genuine consumer need does it satisfy? The average user does not need to win the *Jeopardy!* television game show or defeat world chess champion Garry Kasparov or outmaneuver world Go champion Lee

Se-dol. They do not need to have this full capability built into their personal devices.

I think the way it will happen is that AI will exist in the cloud, in various localized networks and only to a limited extent in the device. The function will be distributed at different points in the ecosystem. AI will have to become faster than it is today and will have to become more customized than it is today to satisfy the needs and desires of individual users, but that is already happening. AI functions will include voice recognition, language and semantic analyses, computer vision, and similar advanced functions. Those functions will know your preferences and make recommendations to you, just as Amazon. com's algorithms already "know" what you have purchased in the past or what other items might naturally go with an item you have in your shopping cart. These systems improve the more you use them.

Mobile PCs also will benefit from the upgrade of cellular telephone networks from their current fourth generation (4G) to the fifth (5G). That will allow for much better natural voice communication capability. I think the mobile PC will become the most powerful device used to integrate different AI functions because they are larger and thus have more room to work with than do mobile phones. That's why I argue that mobile PCs will be the dominant tool that individuals will use to take advantage of the different types of AI that really matter to them.

That, for me, is the ultimate guiding principle. Everything should be based on the needs and desires of the user. We call that customer-centric design. The technology must be for a human purpose. People have to trust it and find it intuitive to work with. We don't want any Hals, the computer in the movie *2001: A Space Odyssey* that came to be the symbol for computers that rage out of control.

I think the human problems that clearly need solving—and can be solved within five years—are the obvious irritants that we all experience in our daily lives. As all the sensors, cameras, and other devices in the Internet of Things improve and as networks become even faster, you will be able to expand the use of cams and other devices to make sure your children are safe at home if you are away or that your home's security has not been breached by criminals. There are some products on the market already, but they are still too difficult for the average user. They need to improve and become second nature to use.

Another human problem: What happens if FedEx or UPS attempts to make a delivery to your home or apartment when you are not there? You can order anything online, but the delivery still takes place in twentieth-century ways. The truck comes. The driver rings the doorbell. You aren't there. The driver leaves a piece of paper telling you to contact one of the delivery services or go to the post office.

The way to solve that problem is that if I have a phone or another mobile device, I can see the deliveryman at my door through a small camera positioned in the right place. My door will have a small delivery door big enough for most packages. I can then remotely open that door for him to make the delivery.

I know from personal experience that there are issues with all these services. My wife wanted to install a camera in our home to watch our dog when we are away. She wanted to see the dog and talk to it.

I bought the necessary equipment, but I was slow to install it. I realized that data from the inside of my home would be going out on a network I was not sure about. I also worry about the ability of other people, even well-meaning people, to take action that would affect my home. If you live in an apartment building, where other units are close by, you don't want your air conditioner to respond to someone else who is at home while you are away. You could return home and find it either too hot or too cold.

Here is how I solved the dog-watching problem. I connected the system and it works perfectly—my wife can watch the dog and even speak and listen to him when we are away. My key decision was to always turn the system off when we are home. I don't want images of us or sounds from us transmitted out to a network I still don't fully trust.

So all these remote capabilities—accepting deliveries, observing children and pets, and controlling climates are not quite ready, but they will be soon. We just need to be mindful of how we use them.

The Twenty-Year View

One of the major issues I've been wrestling with in recent years, and still today, is to find, recruit, and establish the next generation of great engineers and innovators in all of Lenovo's PC labs. As I have said, I believe I have institutionalized the culture and the leadership that we have enjoyed over the decades.

We don't know whether innovation will follow a straight-line projection or undergo some fundamental form of disruption. A disruption would occur, for example, if we learned how to insert needles into the right parts of our brains and those needles either contained computer circuitry or were connected remotely to computers. The brain's communication with the outside world would be enhanced.

If that doesn't work, it might be that computerized devices could be inserted elsewhere in the body, to help a paralyzed arm or leg function normally for example. I was very excited by reports in April 2016 that an American man who had become a quadriplegic five years ago as a result of a diving accident was able to move his arm for the first time because of a tiny device implanted in his brain. He then used his brain signals to move the arm. I'm not an expert on the science of the human body, but I know that doctors also are inserting probes into brains to create small flows of electricity to areas that control Parkinson's disease. The results of this deep brain stimulation have been very positive.

One of the young engineers I have been grooming is Yasumichi Tsukamoto. His family name is difficult to pronounce even for Japanese—the "ts" sound in his last name is too hard. So we simply call him Moto. He is a ThinkPad fanatic and even wears clothing and shoes with black and red accents. He has been in charge of building interdisciplinary teams that attempt to integrate different technologies into the ThinkPad.

Moto can imagine a world I cannot. It is one in which computer displays become ubiquitous. The cost drops so much that there can be all sizes of displays in offices, homes, and public spaces.

In such a world, each of the computer screens would be linked via networks. We simply will identify ourselves to any computer screen by using our fingerprints or a retinal scan, iris recognition, or just facial recognition. We won't need passwords, but we will still have very strong security.

Because our content will be centralized in the cloud, Moto thinks it is possible that wherever we are, we will be able to access the Internet and all our content, whether documents, music, videos, photographs, or anything else. Therefore there will be no need to carry any device, much less two.

He is also very interested in what Amazon is doing with its Echo group of products. These devices appear to be very simple to the user. The user talks to the devices by engaging through Amazon's Alexa. The devices are connected to the Internet, but the user does not have to boot up a machine or enter credit card information. In a very natural way, one can order food items or personal toiletries (or anything else) that Amazon will then deliver. "The technology is user-centric and very simple, but the back end of the system contains major complexity," Moto says. The architecture of systems and networks fascinates him. At his age, I was thinking about one machine. He is thinking about networks.

Ultimately, he has big dreams. "What I'd like to do is create an entirely new segment of products and services," Moto says. "I also want to do something that changes people's lives whether they are at work or at home. I want to help them become more productive in business while making their lives easier. My dream is for ThinkPad to create a new segment, a new user experience," Moto says. Whenever

I hear him and other young engineers talk about their ambitions like this, it makes me very happy.

The Broader Response to Technology

It's obvious to me and many other people that technology has moved ahead so fast, and will continue to move ahead so quickly, that it has far outstripped the ability of governmental, educational, and cultural institutions to understand what has happened, much less respond to it. There are discontinuities. Look at how fast pay telephones and phone booths have disappeared, how people no longer use paper maps, and how people no longer have heavy encyclopedias occupying their bookshelves. No one goes to Blockbuster anymore to rent movies. Not many people use travel agents because it is so easy to book arrangements online. The news media, book publishing, and music worlds have been completely disrupted, and the taxi and hotel industries could be next if Uber and Airbnb continue to succeed.

This is an unscientific opinion, but it seems to me that during previous periods of rapid technological change, such as the Industrial Revolution in the West or the introduction of silk technology into Japan, societies and governments had time to adjust more smoothly to the new capabilities. To me, the changes we are now living through seem more disruptive.

I am sometimes asked whether we technology innovators have some sort of moral responsibility to address the issues that this latest explosion of technology has created. Security and privacy are two of the biggest issues obviously.

I respond by saying that engineers cannot control how the things we invent are used. The case of the drone is a good example. People are now able to make drones that can fly aloft for many hours and cover great distances. There are many positive uses: Amazon is considering using them to deliver its packages. News organizations and government agencies can use drones installed with cameras to survey floods and other natural disasters. But drones can also carry spy cams and even weapons.

It has taken time for governments to determine who has the responsibility for setting rules that apply to drones. Some argue they should be banned from school areas because of their capacity for harming children. They should be banned from airport areas, it seems obvious. But it has taken time for the right regulatory decisions to be made. Policy lagged behind the technology.

Autonomous cars are another example of how technology has run ahead of government's ability to regulate or govern or manage. If an autonomous car causes an accident, who is responsible? There needs to be a debate and a tangible response.

Or take the case of three-dimensional printing. This is an entirely new way of manufacturing things, and it has potential to be misused. Reports have surfaced of people finding specifications for plastic handguns on the Internet and being able to make the guns on their home 3-D printers. Society and government have to figure out a way to respond to that danger.

As a technologist, my responsibility, as I have said repeatedly, is to create technologies that address the needs of people—to try to help improve their lives and to enable something that has not been possible before. It is my job to change the way devices behave so that they have more human interfaces. People need to control each of their devices. These devices should be integrated and intelligent. You know your own style of engagement and the devices should respond to you in the way you want them to—they should be intuitive. The engagement should feel natural. Until now, people have had to learn how to use the machine's interface in order to communicate with it. Now it is the machine's turn to learn the human interface. This has been a dream since the 1960s, but the technologies to actually make it happen are just around the corner.

But everyone who uses these new tools must educate themselves about the dangers as well as the opportunities. For example, I use Facebook. It's very convenient. My own personal policy is that I post my own picture, but I do not post any new images of other people. I also do not post images of buildings or outdoor scenes, and I do not post images of myself, while I am away from home, if those images reveal where I am geographically. I do not know where those images could end up. I do not want anyone's privacy invaded, including my own.

People owe it to themselves to be selective in how they use Facebook or any other new capability. When Facebook first exploded onto the stage, we accepted every friend request, even those from strangers.

Over time, people have learned how to protect their privacy. If they adjust their settings correctly, they will not disclose personal information. They can turn location services on or off depending on whether they want people to know where they are. They can limit their audience.

In all uses of a computer, security software is essential. If someone uses a computer without making sure he or she has very strong protection, whose fault is it if a hacker is able to penetrate their system? The user should have understood the challenges as well as the opportunities. Users have responsibilities to understand the power of the technology in their hands.

Governments, opinion leaders, and educators all must be involved in discussions about how to respond to new technologies—to guide them toward achieving tangible human goals—because this revolution is not anywhere close to being finished.

On a Personal Note

Having been diagnosed with lung cancer six years ago, and undergoing a painful operation that appears to have removed it all, I have had plenty of time to contemplate questions of mortality. I think back to my father. He was just a regular citizen. He was not famous in any way. But without my recognizing it (at least for many years), he taught me how to think and how to solve problems that people face. He helped me define an emotional center. He shaped my behavior. As a result, he was still alive in me long after he was physically gone from this world.

If I have been successful in transferring my knowledge, basic behaviors, and style to my own children and to my young engineers, that's all I ask for. That's how I want to close out my life. I don't care if people remember my name. I don't care whether they recognize that I contributed to their thought processes. If the spirit of my life's work is transferred to the next generation, to my army of innovators, I'll still be there.

Index